# Lecture Notes in Control and Information Sciences

Edited by M. Thoma and A. Wyner

IFIP

87

# Recent Advances in System Modelling and Optimization

Proceedings of the IFIP-WG 7/1 Working Conference
Santiago, Chile, August 27-31, 1984

Edited by
L. Contesse, R. Correa, and A. Weintraub

Springer-Verlag
Berlin Heidelberg GmbH

**Editors**

Luis Contesse B.
Depto. Matemáticas y Ciencias de la Computación
Facultad de Ciencias Físicas y Matemáticas
Universidad de Chile
Casilla 170/3, Correo 3
Santiago, Chile

Rafael Correa F.
Depto. Matemáticas y Ciencias de la Computación
Facultad de Ciencias Físicas y Matemáticas
Universidad de Chile
Cassilla 170/3, Correo 3
Santiago, Chile

Andrés Weintraub P.
Depto. Ingeniería Industrial
Facultad de Ciencias Físicas y Matemáticas
Universidad de Chile
Santiago, Chile

ISBN 978-3-540-17083-9 ISBN 978-3-540-47201-8 (eBook)
DOI 10.1007/978-3-540-47201-8

Library of Congress Cataloging in Publication Data


Originally published by Springer-Verlag Berlin Heidelberg New York in 1986

2161/3020-543210

# PREFACE

These Proceedings include most of the papers presented at the IFIP Working Conference on System Modelling and Optimization held in Santiago, Chile, August 27-31, 1984.

The Conference was attended by scientists of both North and South America. The program offered a view of optimization techniques and models currently in use and under investigation. Major emphasis was on advances in Control System Design, Mathematical Programming Theory and Methods, Combinatorial Optimization and the application of all these techniques to Modelling. In particular, one of the objectives of the meeting was to report on the use of quantitative models in the area of natural resources management and planning. Three relevant areas were chosen: mining, timber and hydrological resources.

The Local Organizing Committee is grateful to Professor A.V.Balakrishnan and Professor E. Polak for the interesting lectures "Kalman Filtering: Theory and Applications" and "Theoretical and Software Aspects of Optimization - Based Control System Design" respectively, delivered during the week proceeding the Conference. The Committee also wishes to thank all the participants at the Conference and the contributors of papers, without whom neither the Conference nor these Proceedings would have been possible.

Finally, the Local Committee is specially grateful to Professor A.V. Balakrishnan, Chairman of the International Programe Committee, for granting to the Facultad de Ciencias Físicas y Matemáticas de la Universidad de Chile, the privilege of hosting this Conference.

Luis Contesse        Rafael Correa        Andrés Weintraub

TABLE OF CONTENTS.

# THE ARMIJO STEP RULE ADAPTED TO GRADIENT PATH ALGORITHMS.

Jorge Amaya
Departamento de Matemáticas y Ciencias de la Computación
Facultad de Ciencias Físicas y Matemáticas
Universidad de Chile
Casilla 170 -Correo 3, Santiago de Chile

ABSTRACT.

In a previous work of the author, the convergence of the algorithm

$$x_{k+1} = x^k(t_k) = x_k - S_k(t_k)\ \nabla f(x_k) \qquad k = 0,1,2,\ldots$$

for the minimization of $f: \mathbb{R}^n \to \mathbb{R}$, was analyzed. Due to the nonlinearity of the matrix $S_k$ the cost of evaluation of $x^k(t)$ for several values of t can be expensive, hence the choice of the first trial $\alpha_k$ is critical.

In this work a procedure for the choice $\alpha_k$ is proposed. This choice is based on the form of the curve $x^k(t)$, $t \geq 0$ and under the convergence hypothesis of the algorithm.

## 1. INTRODUCTION.

The present work is intended as a contribution to the analysis of the step size problem in algorithms of the gradient path type. Two algorithms of this family are specifically analized.

Let $f : \mathbb{R}^n \to \mathbb{R}$ be a $C^2$ function which is to be minimized. Given $x_k \in \mathbb{R}^n$, the gradient path emerging from $x_k$ is defined by the differential equation ([3], [4])

$$\begin{aligned} \dot{x}(t) &= -\nabla f(x(t)) \qquad t \geq 0 \\ x(0) &= x_k \end{aligned} \qquad (1)$$

whose solution, under appropriate hypotheses, converges to a stationary point of f as t approaches infinity.

A linear approximation for f around $x_k$ yields the following approximate solution of (1):

$$x^k(t) = x_k + (e^{-tH(x_k)} - I)\, H(x_k)^{-1}\, g_k,$$

where $H(x_k)$ and $g_k$ respectively denote the hessian matrix and the gradient of f, both evaluated at $x_k$.

Expression (2) suggests and algorithm defined by the iteration formula

$$x_{k+1} = x^k(t_k) = x_k + (e^{-t_k H_k} - I)\, H_k^{-1} g_k, \tag{3}$$

where $H_k$ is a positive definite approximation of $H(x_k)$ and $t_k$ the step size.

The convergence of algorithm (3) has been studied in [1] under a more general formalism characterized by the iteration formula

$$x_{k+1} = x_k - S_k(t_k) g_k, \tag{4}$$

where $S_k : \mathbb{R}^+ \to \mathbb{R}^{nxn}$ is a matrix of class $C^1$. The step size is determined by a criterion of the Armijo type [2] : let $\delta, \beta \in ]0,1[$ and $\alpha_k > 0$; then

$$t_k = \max\{\lambda \mid \lambda = \alpha_k \beta^s, s = 0,1,2.. \; f(x^k(\lambda)) \leq f(x_k) - \lambda\delta < g_k, \check{S}_k(0) g_k >\} \tag{5}$$

An important drawback of algorithm (4) is, in our case, the evaluation of matrix $e^{-tH_k}$ at several values of t in the course of each iteration. When the spectral decomposition of $H_k$ is available this evaluation is simple, but determining in a stable form the system of eigenvalues and eigenvectors of the hessian matrix in each iteration is known to involve a fairly high computational effort. The choice of $\alpha_k$ (first trial in the k-th iteration) is thus cardinal to attaining a low computational cost.

## 2. CHOICE OF $\alpha_k$ .

Since we have assumed $H_k$ to be positive definite, the curve $x^k(t)$ , $t \geq 0$, is an arc emerging from $x_k$ and satisfving $x^k(t) \to x_k - H_k^{-1} g_k$ as $t \to \infty$. In addition, its derivative at the origin is $-g_k$.

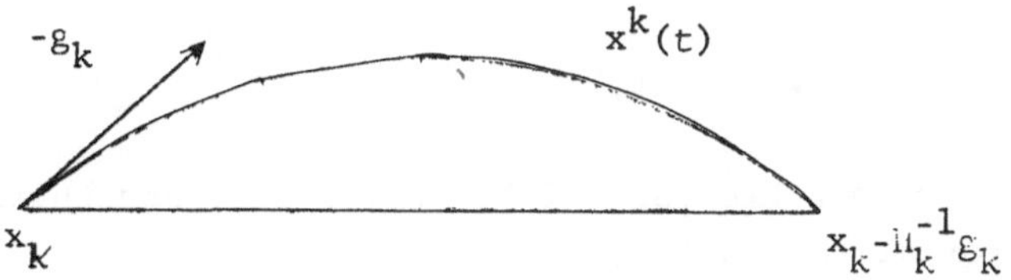


In case that $H_k$ equals $H(x_k)$, a rate of convergence as good as that of Newton's method can be attained by examining in each iteration whether $x^k(\infty)$ satisfies the inequality in (5). Otherwise, the procedure goes on to examine step sizes $\alpha_k \beta^s$ for $s = 0,1,2...$

For a given $t > 0$, it is not in general possible to foresee the location of $x^k(t)$ on the curve, nor is it its distance from $x_k$. Such will depend on the eigenvalues of $H_k$.

We shall deduce a choice of $\alpha_k$ such that the distance between $x^k(\alpha_k)$ and $x_k$ is a certain fraction of the distance between $x^k(\infty)$ and $x_k$. That is, for a given value $\gamma \in ]0,1[$, we impose the condition

$$\frac{||x^k(\alpha_k) - x_k||}{||x^k(\infty) - x_k||} = \gamma \tag{6}$$

If we approximate $e^{-\alpha_k H_k}$ by $I - \alpha_k H_k$, from (3) and (6) we have

$$\frac{||- \alpha_k g_k||}{||H_k^{-1} g_k||} \approx \gamma \tag{7}$$

Hence we propose

$$\alpha_k = \gamma \frac{||H_k^{-1} g_k||}{||g_k||} \tag{8}$$

as a choice for the value of $\alpha_k$.

## 3. CONVERGENCE HYPOTHESES.

Algorithm (4) generates a sequence in $\mathbb{R}^n$ having the property that every accumulation point is a stationary point of f. The hypotheses under which convergence is shown to occur are [1]:

(h1) $L_0 = \{x \in \mathbb{R}^n \ / \ f(x) \leq f(x_0)\}$ is compact

(h2) $S_k(0) = 0 \qquad \forall k = 0,1,2\ldots$

(h3) $S_k(0)$ is symmetric for all $k = 0,1,2,\ldots$, and there exists $m, M > 0$ such that, for all $k = 0,1,2,\ldots$

$$m\,||z||^2 \leq z^T S_k(0)\, z \leq M\, ||z||^2 \quad \forall z \in \mathbb{R}^n \tag{9}$$

(h4) The sequence $\{\dot{S}_k\}$ is equicontinuous at zero, i.e., $\forall \varepsilon > 0\ \exists \eta > 0$ such that

$$t \in ]0,\eta[ \implies ||\dot{S}_k(t) - \dot{S}_k(0)||_\infty \leq \varepsilon \quad \forall k = 0,1,2^{(1)}\ldots \tag{10}$$

(h5) If $\{x_{k_j}\}$ is a subsequence of $\{x_k\}$ converging to a non-stationary point of $f$, then there exists $\nu > 0$ and an integer $\ell > 0$ such that

$$\alpha_{k_j} \geq \nu \qquad \forall j \geq \ell$$

Hypotheses (h2), (h3), and (h4) have been analyzed in [ 1 ]; we shall therefore examine whether the value for $\alpha_k$ chosen according to (8) satisfies (h5). We will first impose the following hypothesis on $H_k$:

(h6) If $\{x_{k_j}\}$ is a subsequence of $\{x_k\}$ converging to a non-stationary point of $f$; then there exists $\ell, m_1, M_1 > 0$ such that for all $j \geq \ell$,

$$m_1\, ||z||^2 \leq z^T H_{k_j}\, z \leq M_1\, ||z||^2 \qquad \forall z \in \mathbb{R}^n \tag{11}$$

From (11) we evidently have, for $j \geq \ell$,

$$\frac{1}{M_1^2}\, ||z||^2 \leq z^T H_{k_j}^{-2}\, z \leq \frac{1}{m_1^2}\, ||z||^2 \qquad \forall z \in \mathbb{R}^n,$$

that is,

$$\frac{1}{M_1^2}\, ||z||^2 \leq ||H_{k_j}^{-1} z||^2 \leq \frac{1}{m_1^2}\, ||z||^2 \qquad \forall z \in \mathbb{R}^n$$

By applying this inequality to $z = g_{k_j}$ we have for $j \geq \ell$

$$\frac{\gamma}{M_1} \leq \alpha_{k_j} \leq \frac{\gamma}{m_1} \tag{12}$$

---

(1) $||\ \ ||_\infty$ denotes the matrix norm $||A||_\infty = \sup_{||x||=1} \{||Ax||\}$

Expression (12) leads directly to a proof that hypothesis (h6) is fulfilled by a choice of $\alpha_k$ according to (8).

An algorithm very similar to (3) is one proposed by Zang [5], which involves expressing matrix $H_k$ in the form of Cholesky's decomposition. In this case the iteration formula is

$$x_{k+1} = x^k(t_k) = x_k + L_k^{-T}(e^{-t_k D_k} - I)\, D_k^{-1} L_k^{-1} g_k \tag{13}$$

where $L_k D_k L_k^T$ is the Cholesky's decomposition of $H_k$. It is clear that $x^k(t) \to x_k - H_k^{-1} g_k$ as $t \to \infty$ and $x^k(0) = -(L_k L_k^T)^{-1} g_k$.

By considering (6) we have

$$\frac{\|L_k^{-T}(e^{-\alpha_k D_k} - I)\, D_k^{-1} L_k^{-1} g_k\|}{\|H_k^{-1} g_k\|} = \gamma$$

which, by using the first order approximation of $e^{-\alpha_k D_k}$, origins the choice

$$\alpha_k = \gamma \frac{\|H_k^{-1} g_k\|}{\|(L_k L_k^T)^{-1} g_k\|} \tag{14}$$

This choice satisfies hypothesis (h5). In fact, from (h6) we have, for $j \geq \ell$,

$$\frac{1}{M_1^2}\|z\|^2 \leq \|H_{k_j}^{-1} z\|^2 \leq \frac{1}{m_1^2}\|z\|^2 \qquad \forall z \in \mathbb{R}^n \tag{15}$$

On the other hand, from lemma 3.1 of [1],

$$\frac{m_1}{M_1}\|z\|^2 \leq z^T (L_{k_j} L_{k_j}^T)^{-1} z \leq \|z\|^2 \qquad \forall z \in \mathbb{R}^n$$

which implies

$$\frac{m_1^2}{M_1^2}\|z\|^2 \leq z^T (L_{k_j} L_{k_j}^T)^{-2} z \leq \|z\|^2 \qquad \forall z \in \mathbb{R}^n$$

or

$$\frac{m_1^2}{M_1^2}\|z\|^2 \leq \|(L_{k_j} L_{k_j}^T)^{-1} z\|^2 \leq \|z\|^2 \qquad \forall z \in \mathbb{R}^n$$

By combining this inequality with (15) we obtain, for $z = g_{k_j}$,

$$\frac{1}{M_1^2}||(L_{k_j}L_{k_j}^T)^{-1}g_{k_j}||^2 \leq ||H_{k_j}^{-1}\ g_{k_j}||^2 \leq \frac{M_1^2}{m_1^4}||(L_{k_j}L_{k_j}^T)^{-1}\ g_{k_j}||^2$$

thus,

$$\frac{\gamma}{M_1} \leq \alpha_{k_j} \leq \gamma\frac{M_1}{m_1^2} \qquad \forall j \geq \ell \tag{16}$$

which completes the proof.

REFERENCES.


[ 1 ] Amaya, J., On the convergence of curvilinear search algorithms in unconstrained optimization, Operations Research Letters. Vol. 4, 1985.

[ 2 ] Armijo,L., Minimization of functions having Lipschitz continuous first partial derivatives, Pacific Journal of Mathematics. 16(1-3) 1966.

[ 3 ] Botsaris, C.A. and D.H. Jacobson., A Newton-type curvilinear search method for optimization, Journal of Mathematical Analysis and Applications. 54(217-229) 1976.

[ 4 ] Vial, J. Ph. and I. Zang., Unconstrained optimization by approximation of the gradient path. Mathematics of Operations Research. 2(3) 1977.

[ 5 ] Zang, I., A new arc algorithm for unconstrained optimization Mathematical Programming. 15(36-52) 1978.

# ON EFFECTIVENESS MEASURES FOR OPTIMAL SEARCH METHODS.

PATRICIO BASSO
UNIVERSIDAD DE CHILE
FACULTAD DE CIENCIAS FISICAS Y MATEMATICAS

Casilla 170 , Correo 3, Santiago de Chile

ABSTRACT.

Classical global problems are subsumed under a single definition, making it possible to formulate unified criteria for measuring the effectiveness of search methods and to handle problems with more than one solution. A number of relevant criteria are proposed for a prescribed class of functions. Explicit expressions of these criteria for the global maximum search problem in the class of Lipschitz-continuous functions are obtained and the so-called optimal one-step methods are determined.

## 1. INTRODUCTION.

A search method is defined by the strategy of choosing the points where a given function will be evaluated in order to estimate some of its global characteristics such as the value of its global maximum, the arguments where this value is attained, or its roots in a given interval. These are what we can call <u>classical global problems</u>.

Since the introduction of the optimal minimax search method of Fibonacci for unimodal functions by Kiefer [7] in 1953, a number of new-methods have been developed for the global maximum search problem. These new-methods are intended to extend Kiefer's results to other classes of functions, to introduce other kind of strategies and to define new criteria for measuring the effectiveness of a strategy. See Basso [2] for references on this subject.

Search methods and optimality criteria developed for the global maximum problem have been applied to the search for a root of a function, the earliest work on this direction being that of Gross and Johnson [6] in 1959. See Basso and Gatica [3] for references.

Two main criteria, called D and W, have been used in the literature for measuring the effectiveness of a search method. Criterion D, introduced by Kiefer, is defined as the length of the shorter interval in which, taking into account the calculated valuesof the function, the required argument can be located.

Criterion W, apparently introduced by Chernousko [ 4 ] but actually called W by Sukharev [10] , is defined as the difference between the global maximum and the discrete maximum obtained with the calculated values of the function, i.e.it is only available for the search of the value of the global maximum.

This paper is motivated by the following considerations:

a) In almost all the cases, a uniqueness hypotheses on the number of solutions of the problem is necessary to obtain meaningful optimal strategies. Hence, neither the problem of determining the set of points where a non-unimodal function reaches its global maximum nor the root-searching problem for functions having several roots can be handled with the criteria used at present.

b) The effectiveness criteria are problem dependent. It would be useful to have a general definition of a global problem with respect to which effectiveness criteria could be defined.

c) Criterion W is of a nature different from that of criterion D. Indeed, while criterion D is an uncertainty measure, criterion W is an error measure for an estimate of the global maximum.

In this paper, we first give a general definition of global problem that considers set-valued maps instead of non-linear functionals so that several-solution cases can be handled. The idea of using set-valued maps for obtaining optimal algorithms was introduced by Traub, Wasilkowski and Wozniakowski [ 12 ].

We then propose some criteria for measuring the effectiveness of a strategy seeking the value of a general global problem. These criteria include a generalization of criterion D and of a criterion introduced by Basso and Gatica [ 3 ] for the root-searching problem, called criterion C. Besides, two new criteria, called criterion A and B, are introduced.

Criterion W cannot be included in this general framework because of its different nature. Nevertheless, we state a relation between criteria W and D that shows that the latter is a useful criterion.

Finally, we obtain explicit expressions of the proposed criteria for the classical global problems in the class of Lipschitz-continuous functions and we determine the so-called optimal one-step methods with respect to each one of the proposed criteria for the global maximum problem.

## 2. FRAMEWORKS AND DEFINITIONS.

Let F be a class of real-valued functions defined on an interval [a,b] and let G denote a set-valued map associating to each $f \in F$ and each $E \subseteq [a,b]$ a subset of $\mathbb{R}$, i.e.

$$G : (f,E) \to G(f,E) \subseteq \mathbb{R}$$

<u>Definition 1</u>. A set-valued map G such that for any partition $x_o = a \leq x_1 < \ldots < x_n \leq b = x_{n+1}$ satisfies the hypotheses

$$\text{(H)} \quad G(f,[a,b]) \subseteq \bigcup_{i=0}^{n} G(f,[x_i,x_{i+1}]), \qquad \forall f \in F$$

will be called a <u>global problem</u> on F. For $E \subseteq [a,b]$ and $f \in F$, the elements of G(f,E) will be called the <u>solutions</u> of the global problem G on E.

Let $E \subseteq [a,b]$ be given and let us define on F the non-linear functional

$$\alpha_E(f) = \sup_{x \in E} f(x),$$

which for E = [a,b] will be simply denoted $\alpha(f)$.

In this paper we deal with the following set-valued maps that, as it is easy to prove, are global problems:

<u>Value-of-the-global-maximum problem</u>: $G_1(f,E) = \{\alpha_E(f)\}$.

<u>Arguments-of-the-global-maximum problem</u>: $G_2(f,E) = \{x \in E / f(x) = \alpha_E(f)\}$.

<u>Roots problem</u>: $G_3(f,E) = \{x \in E / f(x) = 0\}$.

Definition 2. Let G be a global problem on F. A strategy of length n seeking the solutions of G on $E \subseteq [a,b]$ is a map $X_n$ that associates to each $f \in F$ a set of n points in [a,b] at which f(x) will be evaluated in order to estimate the value of G(f,E), i.e.

$$X_n : f \in E \to X_n(f) = (x_1,\ldots,x_n) \in \mathbb{R}^n.$$

Remark 1. The above definition can be generalized by considering, instead of the evaluation of f(x), an information operator. See Traub, Wasilkowski and Wozniakowski [12] and Basso [2] .

Definition 3. A search method for the solutions of a global problem G on $E \subseteq [a,b]$ is a sequence $\{X_n\}$, $n \geq 1$, of strategies.

Definition 4. Let $f \in F$ and let $X_n$ be a strategy seeking the solutions of a global problem G on $E \subseteq [a,b]$. The localization or uncertainty set for the solutions of G on E is the smallest set $L_n$ containing G(f,E) that can be actually determined after the application of strategy $X_n$.

Let $F_n(f)$ be the class of functions of F sharing the same information as $f \in F$ after a given strategy $X_n(f) = (x_1,\ldots,x_n)$ has been applied, i.e.

$$F_n(f) = \{g \in F / g(x_i) = f(x_i),\ i = 1,\ldots,n\}.$$

Hence, as it is easy to prove, the localization set $L_n$ is given by

$$L_n = \bigcup_{g \in F_n(f)} G(g,[a,b]).$$

Let us define the sets:

$$L_{n,i} = \bigcup_{g \in F_n(f)} \{G(g,[a,b]) \cap G(g,\ [x_i,x_{i+1}])\},$$

i.e. the sets of points that are simultaneously solutions of G on [a,b] and on $[x_i,\ x_{i+1}]$ for some $g \in F_n(f)$.

Then, due to hypotheses (H), we have

$$L_n = \bigcup_{i=0}^{n} L_{n,i}.$$

Let us denote by $\ell_i$, $i = 0,\ldots,\ n$, the Lebesgue measure of the sets $L_{n,i}$

and let $\vec{\ell} = (\ell_0, \ldots, \ell_n)$ be the vector of these values. Then, we will define the following criteria for measuring the effectiveness of strategy $X_n$:

Criterion L : $L(X_n, f)$ = the Lebesgue measure of $L_n$.

Criterion D : $D(X_n, f)$ = the diameter of $L_n$.

Criterion A : $A(X_n, f) = ||\vec{\ell}||_\infty = \max_{i=0,\ldots,n} \ell_i$

Criterion B : $B(X_n, f) = ||\vec{\ell}||_1 = \sum_{i=0}^{n} \ell_i$

Criterion C : $C(X_n, f) = ||\vec{\ell}||_2^2 = \sum_{i=0}^{n} \ell_i^2$

Remark 2. Exceptin the paper of Basso and Gatica [3], where criterion C was introduced, only criterion D has been used for the global problems $G_2$ and $G_3$.

Remark 3. Gal and Micchelli [5] used criterion D in the general framework of the optimal search for the value of a functional which, clearly, includes global problem $G_1$. Aside from this, only Basso [2] has used criterion D for $G_1$.

Remark 4. Criterion D is meaningless for the obtainment of the optimal strategies unless a uniqueness hypotheses on the number of solutions is made. For instance, D-optimal strategies for the global problem $G_2$ and $G_3$ are meaningless in the class of Lipschitz-continuous functions as was shown by Sukharev [10], [11].

## 3.- SOME REMARKS ON CRITERION W.

As we pointed out in remark 3, criterion D has not been used for global problem $G_1$. Instead, most authors have used a criterion called W that is defined as follows:

$$W(X_n, f) = \alpha(f) - \alpha_n(f),$$

where $\alpha_n(f)$ is the discrete maximum

$$\alpha_n(f) = \max_{i=1,\ldots,n} f(x_i)$$

Criterion W and the criteria defined in the previous section are of a different nature because the former is an error measure for the estimate $\alpha_n(f)$ while the latter are related to the uncertainty set of $\alpha(f)$.

Criteria W and D are used for the obtainment of the minimax optimal strategies where the quantities

$$W(X_n) = \sup_{f \in F} W(X_n, f)$$

and

$$D(X_n) = \sup_{f \in F} D(X_n, f),$$

called the <u>result guaranteed by strategy $X_n$</u> on F, area utilized.

The following lemma allows comparing these two quantities:

<u>Lemma</u>. Let $X_n$ be a strategy of length n on F and let $f_n^-$ be the lower envelope of $F_n(f)$, i.e.

$$f_n^-(x) = \inf_{g \in F_n(f)} g(x).$$

Then,

i) $D(X_n) \leq W(X_n)$

ii) If $\alpha(f_n^-) = \alpha_n(f)$ for all $f \in F$, then $D(X_n) = W(X_n)$.

<u>Remark 5</u>. Condition (ii) in the preceding Lemma is satisfied in every class of functions at which, criterion W has been applied. Hence, in these classes criteria D and W produce the same minimax optimal strategies.

<u>Remark 6.</u> According to the definition of criteria D and W, the following sharpest bound can be established:

$$|\alpha(f) - \alpha_n(f)| \leq W(X_n) \qquad \forall f \in F$$

$$|\alpha(f) - \alpha(f_n^-)| \leq D(X_n) \qquad \forall f \in F$$

$$|\alpha(f) - \hat{\alpha}_n(f)| \leq \frac{1}{2} D(X_n) \qquad \forall f \in F$$

where the estimate $\hat{\alpha}_n$ is defined as

$$\hat{\alpha}_n(f) = \frac{1}{2} (\alpha(f_n^+) + \alpha(f_n^-))$$

with $f_n^+$ the upper envelope of $F_n(f)$.

Taking into account the foregoing remarks and the results of the lemma, we conclude that it seems better to use criterion D rather than criterion W for the search of the value of the global maximum.

## 4. EFFECTIVENESS CRITERIA IN $LIP_K(a,b)$.

In this section we will utilize criteria L,D,A,B and C for measuring the effectiveness of a strategy seeking the solutions of the global problems $G_1$, $G_2$ and $G_3$ in the class of the Lipschitz-continuous functions.

Let $K \geq 0$ and let $F = Lip_K(a,b)$ be the class of functions f such that

$$|f(x) - f(y)| \leq K\ |x-y| \qquad \forall x,y\ [a,b]$$

For $f \in F$ let $X_n(f) = (x_1,\dots,x_n)$ be a strategy seeking the solutions of a global problem $G \in \{G_1,G_2,G_3\}$ and let us assume, without loss of generality, that $x_o = a \leq x_1 < \dots < x_n \leq b = x_{n+1}$. Hence, as it is easy to prove, the class $F_n(f)$ has upper and lower envelopes given by

$$f_n^+(x) = \min_{i=1,\dots,n} \{f(x_i) + K|x-x_i|\},$$

$$f_n^-(x) = \max_{i=1,\dots,n} \{f(x_i) - K|x-x_i|\}.$$

Let us define the local maxima

$$M_j = \max_{x \in [x_j, x_{j+1}]} f_n^+(x), \qquad j = 0,\dots,n.$$

Hence, we have

$$\alpha(f_n^+) = \max_{j=0,\dots,n} M_j .$$

On the other hand, since $f_n^-$ is a piecewise convex function, we get

$$\alpha(f_n^-) = \alpha_n(f) = \max_{i=1,\dots,n} f(x_i).$$

For the sake of shortness we give, without proof, the final expressions of the different criteria for each one of the global problems $G_1$, $G_2$ and $G_3$.

Problems $G_1$: the value of the global maximum.

$$L(X_n,f) = D(X_n,f) = A(X_n,f) = \alpha(f_n^+) - \alpha_n(f),$$

$$B(X_n,f) = \sum_{j=0}^{n} (M_j - \alpha_n(f))_+,$$

$$C(X_n,f) = \sum_{j=0}^{n} (M_j - \alpha_n(f))_+^2,$$

where, for $r \in \mathbb{R}$, we define $(r)_+ = \max(0,r)$.

It is worth remarking that since the condition $\alpha(f_n^-) = \alpha_n(f)$ holds for all $f \in F$ we have, according to the lemma of section 2, that $W(X_n) = D(X_n)$.

Problem $G_2$: the arguments of the global maximum.

By defining the quantities:

$$a_j = \begin{cases} a & \text{for } j = 0, \\ x_j + \dfrac{\alpha_n(f) - f(x_j)}{K} & \text{for } j = 1,\dots, n, \end{cases}$$

$$b_j = \begin{cases} x_{j+1} - \dfrac{\alpha_n(f) - f(x_{j+1})}{K} & \text{for } j = 0,\dots,n-1, \\ b & \text{for } j = n, \end{cases}$$

$$\hat{a}_n = \min\{a_j / a_j \leq b_j,\ j = 0,\dots,n\},$$

$$\hat{b}_n = \max\{b_j / a_j \leq b_j,\ j = 0,\dots,n\},$$

the required expressions are the following

$$D(X_n,f) = \hat{b}_n - \hat{a}_n,$$

$$L(X_n,f) = B(X_n,f) = \sum_{j=0}^{n} (b_j-a_j)_+,$$

$$A(X_n,f) = \max_{j=0,\ldots,n} (b_j-a_j)_+,$$

$$C(X_n,f) = \sum_{j=0}^{n} (b_j-a_j)_+^2.$$

Remark 7. The intervals $[a_i,b_i]_+$ correspond to the local localization sets $L_{n,i}$ utilized in the definition of the different criteria. The notation $[a_i,b_i]_+$ is used to indicate that $[a_i,b_i]_+ = \phi$ if $b_i < a_i$.

Problem $G_3$ : the roots.

The final expressions of the different criteria in this case are formally the same as those given for global problem $G_2$ but with the points $a_j$ and $b_j$, $j = 0,\ldots,n$ defined by

$$a_j = \begin{cases} a \text{ for } j = 0 \\ x_j + \dfrac{|f(x_j)|}{K} \text{ for } j = 1,\ldots,n. \end{cases}$$

$$b_j = \begin{cases} x_{j+1} - \dfrac{|f(x_{j+1})|}{K} \text{ for } j = 0,\ldots,n-1 \\ b \text{ for } j = n. \end{cases}$$

## 5. OPTIMAL ONE-STEP METHODS FOR $G_1$.

After a criterion for measuring the effectiveness of strategies seeking the solutions of a given global problem has been chosen, one can state the problem of determining the optimal strategy with respect to this criterion.

For this problem to be well-posed, we must establish an optimality criterion and the class of strategies within which the optimal strategy will be searched.

Three main classes of strategies have been used in the literature: passive, sequential and block-sequential. See Basso [2] .

With regard to the optimality criterion, Kiefer's minimax optimality criterion has been used, i.e. an optimal strategy is one which minimizes, among all the strategies of the same kind, the maximum value of the effectiveness criterion on the class of functions being analyzed. Hence, an optimal strategy is one which produces the best guaranteed result in a given class of function, among all the strategies of the same type.

A class of optimal search methods, called optimal one-step methods, was introduce by Chernousko [4]. In an optimal one-step method each new evaluation point is chosen as a minimax optimal strategy of length one on $F_n(f)$ where for n = 0 we take $F_o(f) = F$; i.e. we will chose the next evaluation point so as to minimize the maximum possible value of the effectiveness criterion on the class $F_n(f)$.

If Δ is an effectiveness criterion, the result guaranteed by a strategy $x \in [a,b]$ over the whole class $F_n(f)$ is the worst value of $\Delta(x,g)$ with respect to all the possible functions $g \in F_n(f)$, i.e.

$$\Delta(x) = \sup_{g \in F_n(f)} \Delta(x,g)$$

The optimal one-step strategy on $F_n(f)$ is defined as one which provides the best guaranteed result among all the strategies $x \in [a,b]$, i.e. it is any $x_{n+1} \in [a,b]$ solution of the problem:

$$r_{n+1} = \Delta(x_{n+1}) = \inf_{x \in [a,b]} \Delta(x) = \inf_{x \in [a,b]} \sup_{g \in F_n(f)} \Delta(x,g)$$

Remark 8. If for $x \in [a,b]$, $X_{n+1}$ denotes the strategy of length n+1 defined on F by $X_{n+1}(g) = (x_1,\dots,x_n,x)$; then $\Delta(X_{n+1},g)$ is the Δ-effectiveness of strategy $X_{n+1}$ for $g \in F$ while, in the definition above, the quantity $\Delta(x,g)$ represents the Δ-effectiveness of strategy $X(g)=(x)$ for $g \in F_n(f)$. It is easily shown that for criteria L,D,A,B and C the relation $\Delta(X_{n+1},g) = \Delta(x,g)$ holds for all $g \in F_n(f)$ and all $x \in [a,b]$. Nevertheless, for criterion W we only have $W(X_{n+1},g) \leq W(x,g)$ with equality if and only if $g(x) \geq \alpha_n(f)$.

In this section we will give, without proof, the optimal one-step methods on $F = \mathrm{Lip}_K(a,b)$ for problem $G_1$ with respect to each one of the criteria defined in the second section.

D-optimal one-step method.

Let us define the functions

$$h_j^{(1)}(x) = \begin{cases} \alpha_n(f) + K(x-a) & \text{if } j = 0 \\ K\dfrac{x-x_j}{2} + \dfrac{f(x_j)+\alpha_n(f)}{2} & \text{if } j = 1,\dots,n \end{cases}$$

$$h_j^{(2)}(x) = \begin{cases} K\dfrac{x_{j+1}-x}{2} + \dfrac{\alpha_n(f)+f(x_{j+1})}{2} & \text{if } j = 0,\dots,n-1 \\ \alpha_n(f) + K(b-x) & \text{if } j = n \end{cases}$$

$$h_j(x) = \max\,(h_j^{(1)}(x),\ h_j^{(2)}(x)) \quad j = 0,\dots,n$$

and the quantities

$$M_j^{(3)} = \max_{\substack{i=0,\dots,n \\ i\neq j}} M_i \qquad j = 0,\dots,n$$

where $M_j$, $j = 0,\dots,n$, are the local maxima previously defined.

Since $h_j$ is a convex strictly unimodal function, there exists a unique point $\bar{x}_j \in [x_j, x_{j+1}]$ at which $h_j$ attains its minimum. After minimizing we obtain.

$$\bar{x}_j = \begin{cases} \dfrac{2a+x_1}{3} - \dfrac{\alpha_n(f)-f(x_1)}{3K} & \text{if } j = 0 \\ \dfrac{x_j+x_{j+1}}{2} + \dfrac{f(x_{j+1})-f(x_j)}{2K} & \text{if } j = 1,\dots,n-1. \\ \dfrac{2b+x_n}{3} + \dfrac{\alpha_n(f)-f(x_n)}{3K} & \text{if } j = n. \end{cases}$$

The D-optimal one-step method is defined as follows

1 ) $x_1 = \dfrac{a+b}{2}$

2 ) If $f(x)$ has been evaluated at the points $x_1,\dots,x_n \in [a,b]$ then:

i) If $f_n^+$ attains its global maximum at two or more points in$[a,b]$ then $x_{n+1}$ may be any point in $[a,b]$ and the guaranteed results is $r_{n+1} = D(X_n, f)$.

ii) If $f_n^+$ attains its global maximum at a single point located in the interval $[x_j, x_{j+1}]$ then, if $h_j(\bar{x}_j) \geq M_j^{(3)}$ then $x_{n+1} = \bar{x}_j$;

else $x_{n+1}$ may be any point in the optimal strategies set $\{x \in [x_j, x_{j+1}] / h_j(x) \leq M_j^{(3)}\}$ . In either case the guaranteed result is $r_{n+1} = \max (h_j(\bar{x}_j), M_j^{(3)}) - \alpha_n(f)$.

Remark 9. The condition (i) implies that whatever the chosen point $x_{n+1}$ is, we cannot guarantee a better result than that already obtained with the previous information. Indeed, this is a weakness of criterion D because it cannot distinguish among different strategies when $f_n^+$ has several points of global maximum.

The previously stated results show that the optimal one-step strategy is not necessarily unique; not even in the case that $f_n^+$ has a unique point of global maximum. Nevertheless, it is useful to choose $x_{n+1}$ locally optimal. Hence we propose to choose, as one of the possible optimal strategies, a point $\bar{x}_j$ corresponding to any one of the interval $[x_j, x_{j+1}]$ where $f_n^+$ attains its global maximum. This selection of $x_{n+1}$ allows us to reduce the value of $r_{n+1}$ in a finite number of steps.

Remark 10. Exceptin the first and the last intervals, the points $\bar{x}_j$ are the points where $f_n^+$ attains its local maxima, i.e. $f_n^+(\bar{x}_j) = M_j$, $j = 1, \ldots, n-1$. For $j = 0$, $f_n^+$ reaches its local maximum at point a while for $j = n$ it does at point b. The idea of choosing as $x_{n+1}$ any point where $f_n^+$ attains its global maximum was independently introduced by Shuber [9] and Piyavskii [8] . Nevertheless, this way of choosing $x_{n+1}$ compels to choose a and b as the second and third evaluation points thus producing a worse guaranteed result.

Remark 11. In [1] we analyse some weaknesses of the Shubert-Piyavskii's method and we propose some block-sequential strategies to overcome them. These results can be as well applied to the optimal-one step method here obtained.

B-optimal one-step method.

After carrying out the corresponding calculation it is shown that the B-optimal one-step method is defined as follows:

1) $x_1 = \frac{a+b}{2}$

2) $x_2 = a$ or $x_2 = b$.

3) If $x_2 = a$ then $x_3 = b$; else $x_3 = a$.

4) If f(x) has been evaluated at the points $x_1, \ldots, x_n \in [a,b]$, $n \geq 3$, then $x_{n+1}$ can be arbitrary chosen on [a,b] and $r_{n+1} = B(X_n, f)$. Hence, for $n \geq 3$ we cannot guarantee a better result than that obtained with the first n evaluations of the function; hence this is a meaningless criterion for problem $G_1$ on the class $F=Lip_k(a,b)$.

C-optimal one-step method.

The C-optimal strategy is defined in terms of the points:

$$\hat{x}_j = \begin{cases} \dfrac{4a+x_1}{5} - \dfrac{\alpha_n(f) - f(x_1)}{5K} & \text{if } j = 0 \\[2ex] \dfrac{x_j+x_{j+1}}{2} + \dfrac{f(x_{j+1})-f(x_j)}{2K} & \text{if } j = 1,\ldots,n-1. \\[2ex] \dfrac{4b+x_n}{5} - \dfrac{\alpha_n(f) - f(x_n)}{5K} & \text{if } j = n. \end{cases}$$

as follows:

1) $x_1 = \frac{a+b}{2}$

2) If f has been evaluated at the points $x_1, \ldots, x_n \in [a,b]$ then $x_{n+1}$ is any one of the points $x_j$ corresponding to one of the intervals $[\hat{x}_j, x_{j+1}]$ where $f_n^+$ attains its global maximum. The guaranteed result is

$$r_{n+1} = C(X_n, f) - \frac{1}{2} \max_{j=0,\ldots,n} (M_j - \alpha_n(f))_+^2.$$

Remark 12. In this case there are at most n+1 optimal strategies; this is an advantage with respect to criterion D. Besides, the C-optimal strategy coincides with our proposal of D-optimal strategy except for $j = 0$ and $j = n$.

## 6. OPTIMAL ONE STEP METHODS FOR $G_2$.

The optimal one-step methods for the problem $G_2$ will be described in terms of the intervals $[a_j, b_j]$ that corresponds to the sets $L_{n,i}$ introduced in section 2. The expressions of $a_j$ and $b_j$ are those given in section 4 for problem $G_2$.

For problem $G_2$, criteria L,D and B are meaningless for the obtainment of optimal strategies because whatever the choice of $x_{n+1}$ be, we cannot guarantee a better result than that already obtained with the first n evaluations of the function.

A-optimal one-step strategy.

The A-optimal strategy is defined in terms of the quantities

$$\theta_j = \max_{\substack{i=0,\dots,n \\ i\neq j}} (b_i - a_i)_+, \quad j = 0,\dots,n$$

as follows:

1) $x_1 = \frac{a+b}{2}$

2) If f has been evaluated at the points $x_1,\dots,x_n \in [a,b]$ then
   i) If $f_n^+$ attains its global maximum at two or more points in [a,b] then $x_{n+1}$ may be any point in [a,b] and the guaranteed result is $r_{n+1} = A(X_n,f)$.

   ii) If $f_n^+$ attains its global maximum at a single point located in the inverval $[x_j,x_{j+1}]$ then if $b_j - a_j \leq 2\theta_j$ then $x_{n+1} = \frac{a_j+b_j}{2}$ ; else $x_{n+1}$ may be any point of the optimal strategies set $\{x\in[a_j,b_j] / \max(x-a_j,b_j-x) \leq \theta_j\}$. In any case the guaranteed result is:

$$r_{n+1} = \max\left(\frac{b_j - a_j}{2}, \theta_j\right).$$

Remark 13. As it was pointed out at section 4, criteria A and D are equivalent for problem $G_1$, i.e. $D(X_n,f) = A(X_n,f)$. Thus, since for $j=1,\dots,n-1$ the middle point of the interval $[a_j,b_j]$ is the point where $f_n^+$ attains its global maximum, the A-optimal one-step strategies for problems $G_1$ and $G_2$ are exactly the same excepting if $j = 0$ or $j = n$. Hence, we conclude that it seems better to speak of criterion A instead of criterion D in both problems.

C-optimal one-step method.

1) $x_1 = \frac{a+b}{2}$

2) If f has been evaluated at the points $x_1,\dots,x_n \in [a,b]$ then $x_{n+1}$ is any one of the middle points of the intervals $[a_j,b_j]$ of maximum length and the guaranteed result is $r_{n+1} = \max((b_j - a_j)/2, \theta_j)$

## 7. FINAL REMAKS.

We have introduced five criteria for measuring the effectiveness of a strategy seeking the value of a global problem. One of them, criterion D, is an extension of a criterion already used in the literature. Besides, we have shown that a second widely used criterion, namely criterion W, can be replaced by criterion D when optimal strategies for the search of the value of the global maximum are analysed. On the other hand, in the class of the Lipschitz-continuous functions only criterion A and C can be used for both global problems $G_1$ and $G_2$. Finally, we have shown that only criterion C is well defined in any situation i.e. it is the only one to produce a finite number of optimal strategies whatever the behavior of the upper envelope $f_n^+$ be.

## REFERENCES.


[ 1 ] Basso P., Iterative Methods for the localization of the global maximum, SIAM J. Numer. Anal. 19,4(82),pp.781-792.

[ 2 ] Basso,P., Optimal search for the global maximum of functions with bounded semi norm. To appear in SIAM J. Numer. Anal.

[ 3 ] Basso,P. and Gatica G., Optimal bisection type method for functions having several roots. Submitted for publication.

[ 4 ] Chernousko, F.L., On optimal search for the minimum of a convex function, U.S.S.R. Comput. Math. and Math. Phys. 10,6(70), pp. 20-33.

[ 5 ] Gal,S. and Micchelli, C.A., Optimal sequential and non-sequential procedures for evaluating a functional. Applicable Anal., 10,2(80), pp. 105-120.

[ 6 ] Gross,O. and Johnson, S., Sequential minimax search for the zero of a convex function, Math. Tables and other Aids to Comp. 13(59), pp.51.

[ 7 ] Kiefer , J., Sequential minimax search for a maximum, Proc. Amer. Math. Soc. 4(53),pp.503-506.

[ 8 ] Piyavskii, S.A., An algorithm for finding the absolute extremum of a function, U.S.S.R. Comput. Math. and Math. Phys. 12 (1972),pp. 57-67.

[ 9 ] Shubert B.O., A sequential method seeking the global maximum of a function, SIAM J. Numer. Anal.,9,3(72), 379-388.

[10 ] Sukharev A.G., Optimal strategies of the search for an extremum, U.S.S.R. Comput. Math. and Math. Phys., 11(71),pp.119-137.

[11 ] Sukharev, A.G., Optimal search for the roots of a function satisfying a Lipschitz condition, U.S.S.R. Comput. Math. and Math. Phys. 16,1(76), pp.17-26.

[12] Traub,J.F., Wasilkowski and Wozniakowski,H., Information, Uncertainly. Complexity, Addison-Wesley Publ. Co., 1983.

THE USE OF INTERACTIVE COMPUTING FOR VEHICLE ROUTEING

João José B. Bechara and Roberto D. Galvão
Esso Brasileira de Petróleo S.A. and
Instituto Nacional de Tecnologia
Av. Venezuela 82, sala 604
20081 Rio de Janeiro, Brazil

ABSTRACT

The vehicle routeing problem (VRP) seeks to obtain routes for a fleet of vehicles with capacity constraints, travelling in closed circuits from a central depot and delivering defined amounts of goods to specific points. The optimization criterion most often used is the minimization of the total distance travelled.

In this paper the use of interactive computing for vehicle routeing is examined. The advantages of using a data base for problem formulation are also emphasized. Finally a recently developed interactive routeing system is described, together with its use to solve a sample problem.

## 1. INTRODUCTION

Examined from a broad viewpoint, the minimization of distribution costs from a central depot to a set of geographically dispersed customers should involve both locating the depot and finding routes for the delivery vehicles. The vehicle routeing problem (VRP) is a subproblem of this formulation, that deserves being considered separately for its application to many practical situations.

The VRP has as objective, in its classical form, the design of routes for a fleet of vehicles compatible with the problem constraints, satisfying the existing demand at minimum cost. The minimization of the total distance travelled is the optimization criterion most often used.

Originally formulated by Dantzig and Ramser [5], the VRP has been intensively studied since then. The existing literature on the subject is extensive, including several formulations of the problem, different methods of solution, literature surveys and comparative evaluations of solution methods.

The objective of this paper is not to write about the VRP, but to analyse the use of interactive computing to solve the problem. It is thus assumed that the reader has some previous knowledge about the subject. The interested reader is referred to detailed reviews of the VRP, for example Bechara [1], Belhot [2], Bodin et alii [3], Eilon et alii [6], Galvão [7], Golden et alii [8], Mole [9] and Watson-Gandy and Foulds [12].

In the present paper the use of interactive computing for the VRP is examined, within a methodology that includes the utilization of heuristic algorithms previously developed for the problem. The advantages of using a data base for problem definition and post-optimization analysis are also emphasized. Finally SIR - a recently developed interactive computer system for the VRP is described, together with an example of its use to solve a sample problem.

## 2. INTERACTIVE COMPUTING AND THE VRP

Given recent developments related to on-line applications, a detailed study of the feasibility of using this technology to aid in the solution of complex problems is always advisable. This type of approach is a powerful alternative to systems based on batch operation, and lead to the development of decision support systems. The natural evolution of such applications passed through the inclusion, in the procedure, of some kind of optimization algorithm. Because of the speed with which well designed heuristics provide good solutions to complex problems, heuristic algorithms are often incorporated into interactive computing systems.

Considering now specifically the VRP, recently published material [10,11] report the use of interactive computing for vehicle routeing. In [10] a trial by a large company of a commercially available system is reported, while [11] describes in some detail a package that gives great emphasis to the experience of human schedulers. In both papers a comparison is made between results obtained from interactive packages and from classical vehicle routeing algorithms programmed for batch operation. The main conclusion is that the use of a suitable interactive routeing package can provide good results, in many cases more satisfactory than those obtained by other methods.

In interactive routeing systems, many details relative to clients and vehicles must be made available both to the user and to the system optimizer. If this information can be obtained from a reliable data base, problem formulation and man-machine interaction can be greatly simplified. An interactive package should thus include an on-line data base manager specifically developed for the application.

It is possible therefore to define for the VRP an interactive routeing package composed of 3 basic integrated modules with the following functions:

1. Data Base Manager - responsible for the interactive management of system information, simplifying the tasks of problem formulation and post-optimization analysis.
2. Heuristic Algorithm(s) - capable of suggesting solutions to the problems formulated interactively.
3. Interactive Planning Interface - That allows post-optimization analysis, with interactive access to the data base, through which the user can obtain alternative solutions with minimum effort. The interactive planning interface allows the human user to:
    a. enter his own solution and try to improve it;
    b. verify how his solution compares with those produced by the system optimizer;
    c. improve solutions obtained through the heuristic algorithms.

In Figure 1 the basic structure of the system is presented. A graphical module can be added to the scheme, giving a graphical capability to the system that, among other functions, can facilitate man-machine interaction. It is important to note that the basic principles discussed above can be applied to planning systems of different nature, their use being by no means restricted to vehicle routeing.

## 3. SPECIFICATION OF SIR

This part presents SIR - The interactive routeing system developed in [1]. It discusses the geographical location of clients, programmed orders, the heuristic algorithm available in the system and the system's interactive planning facilities. It finally gives a brief general de scription of SIR.

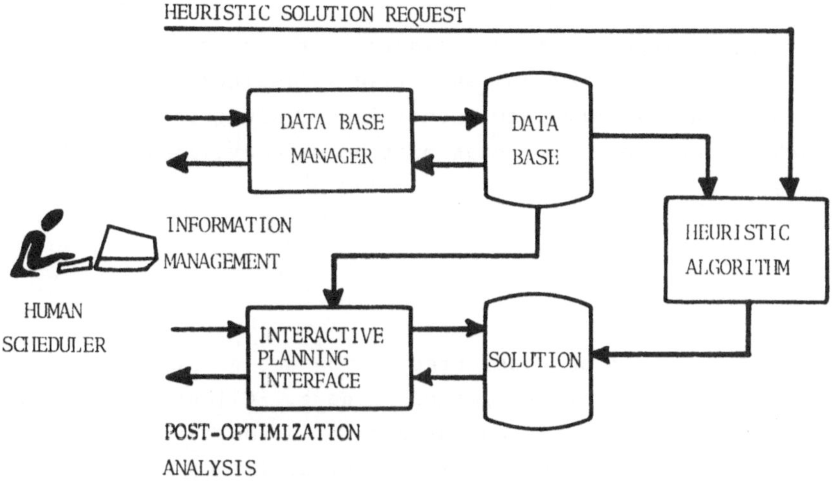


Figure 1 - Interactive Routeing System: Basic Structure

3.1 - Geographical Location of Clients

In the delivery routeing process, the geographical location is the client's most important information. Although the use of (x; y) coordinates facilitates the description of clients' location to the data base, their use for the definition of distances between clients can jeopardize the quality of the solution. This is due to the fact that the real distance between two clients can be significantly different from the Euclidean distance calculated from the (x; y) coordinates.

An alternative would be to store in the data base the real distance between each pair of clients. This approach can however lead to an extremely high number of distances to be stored; for a symmetric distance matrix, n clients define

$$n_d = (n + 1)\, n / 2$$

distances between clients.

The storage problem can be overcome by grouping clients in delivery zones. A delivery zone is a geographical area within which travel time (or distance) between consecutive deliveries is negligible. For better accuracy, however, an average travel time (distance) between clients within each zone may be specified. In this case, for every pair of deliveries within a zone, this average travel time (distance) is added to the total travel time (distance) for each route under consideration.

SIR uses real distances between clients, with the clients grouped in delivery zones. In the client data file it is therefore necessary to indicate to which delivery zone each client belongs. The distance between each pair of delivery zones must be made available to SIR's distance file.

### 3.2 - Programmed Orders

An important aspect to be considered in the design of a vehicle routeing system is the use of programmed orders. Clients with regular demand for a product should be encouraged to authorize deleveries in a programmed basis, instead of placing an order for each delivery. A client's delivery programme should meet its needs for a certain period (e. g. a week or a month), which can aid route schedulers to avoid unbalanced delivery schedules.

SIR allows each client to store programmed orders in its data base. A client with programmed orders may, of course, place additional orders. When the system is run both programmed and not programmed orders are considered in the design of the routes, with the programmed orders recovered from the system's data base.

### 3.3 - The Heuristic Algorithm

As shown in Figure 1, SIR incorporates a module for generating delivery routes, which is done through the use of a heuristic algorithm. The algorithm used by SIR is the savings method of Clarke and Wright [4], in its multiple version.

The idea behind the savings method of Clarke and Wright is as follows [6]. Suppose a depot must supply n customers and that enough vehicles are available, so that each customer can be supplied individually by one vehicle. The total distribution cost is given by

$$2 \sum_{j=1}^{n} C_{oj} ,$$

where $C_{oj}$ is the cost of supplying customer j from the depot o. Now suppose that two customers i and j are linked together so that they are supplied by one vehicle on one route. This eliminates one vehicle route and also reduces the total cost by $S_{ij} = C_{oi} + C_{oj} - C_{ij}$. The quantity $S_{ij}$ is called the saving of link (ij).

The savings method of Clarke and Wright (in its multiple version) can be described as follows [4]:

(a) Calculate the savings $S_{ij}$ for all pairs of customers ij.

(b) Arrange the savings in descending order of magnitude.

(c) Starting from the top of the list, do the following:

  (i) If making a given link results in a feasible route according to the constraints of the problem, then add this link to the solution; if not, reject the link;

  (ii) Try the next link in the list and repeat (i) until the list is exhausted.

(d) The links that have been selected form the solution to the problem.

According to Webb [13], the results produced by the algorithm of Clarke and Wright compare favourably with alternative savings methods developped for the VRP. Although it can be argued that algorithms based on different principles may produce solutions of better quality, the algorithm of Clarke and Wright is computationally very fast, an important feature for an interactive planning system.

It should be noted that given the modular structure of SIR, alternative routeing algorithms may be easily incorporated into the system.

## 3.4 - Interactive Planning Facilities

The system's Interactive Planning Interface gives the user flexibility to enter his own routes and to alter solutions produced by the system, with interactive access to SIR's data base. Once the routes are defined (by the user or through the heuristic algorithm), the following variables are automatically calculated by the system:

(i) For each vehicle:

- load assigned;
- slack capacity;
- total distance travelled.

(ii) For the fleet:
- total capacity;
- total load assigned;
- total slack capacity;
- total distance travelled.

In order to produce alternative solutions, the user must be capable of, for example, inserting deliveries into a route, deleting deliveries from routes, exchanging deliveries between routes. SIR provides the user with 6 options to interactively alter routes:

(a) Move a client from one vehicle route to another, to a specific position in the new route;
(b) Move a client as in (a), but seeking sequence optimization in the new route;
(c) Delete a client from a route;
(d) Exchange 2 clients between 2 routes;
(e) Exchange 2 clients as in (d), but seeking sequence optimization in the 2 routes;
(f) Insert orders / Alter demands.

### 3.5 - General Description of the System

SIR is a computer application developped in MANTIS, a high level user-oriented language. The system includes 3 basic integrated modules (data base manager, heuristic algorithm and interactive planning interface), with its data base composed of 5 files:

- Clients (client information);
- Fleet (vehicle information);
- Distances (distances between delivery zones);
- Orders (programmed orders information);
- Routes (problem solutions).

Figure 2 show SIR's structure and data flow.

## 4. AN ILLUSTRATIVE EXAMPLE

In order to illustrate the use of the system, the following example is considered. An industrial firm has, for a given working day,a delivery programme to 22 clients, to be supplied from a central depot. The de-

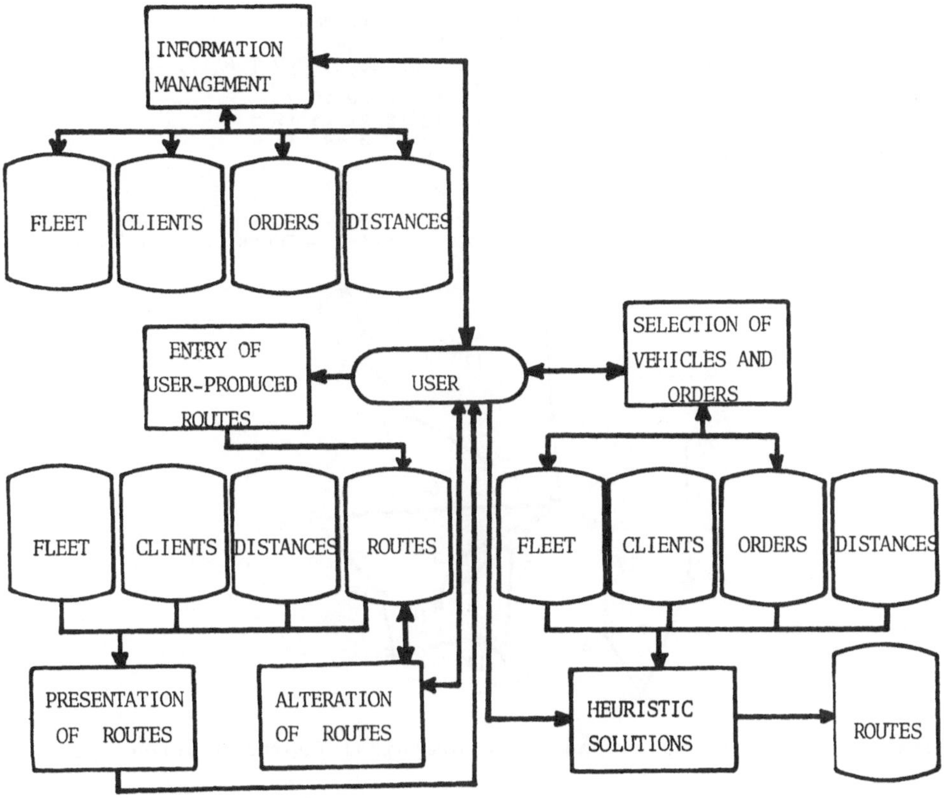


Figure 2 - SIR's Structure and Data Flow

mands of these clients (numbered 1 through 22) are, respectively: 125, 785, 660, 500, 300, 180, 350, 150, 1200, 4000, 400, 1300, 250, 500,150, 100, 250, 130, 600, 500, 175 and 90 kg. The delivery fleet is composed of vehicles of same capacity (4500 kg each), and the maximum distance travelled by a vehicle in a working day cannot exceed 240 km.

Figure 3 shows the heuristic solution obtained for the problem(in SIR's standard format), together with its graphical representation. (The distances shown in Figures 3, 4 and 5 are multiplied by a factor of 10. The headings in the corresponding computer print-outs are in Portuguese.)

An attempt to improve the solution could consider, for example, to transfer client 5 from route (vehicle) 2 to route 4. The system, responding to user request, inserts client 5 in a position in route 4 that mini

DATA: 24/05/84 SISTEMA INTERATIVO DE ROTEAMENTO HORA: 12:20:24

APRESENTACAO DAS ROTAS

| VEICULO | CARGA TOTAL | CAPAC. VEICULO | FOLGA CAPAC. | DIST. TOTAL | ROTA 01 | 02 | 03 | 04 | 05 | 06 | 07 | 08 | 09 | 10 | 11 | 12 |
|---|---|---|---|---|---|---|---|---|---|---|---|---|---|---|---|---|
| 1 | 1625 | 4500 | 2875 | 1783 | 1 | 11 | 16 | 20 | 15 | 7 | | | | | | |
| 2 | 3185 | 4500 | 1315 | 1675 | 6 | 12 | 18 | 21 | 14 | 19 | 5 | | | | | |
| 3 | 3635 | 4500 | 865 | 1477 | 8 | 22 | 2 | 3 | 4 | 17 | 9 | | | | | |
| 4 | 4250 | 4500 | 250 | 766 | 10 | 13 | | | | | | | | | | |
| TOTAIS: | 12695 | 18000 | 5305 | 5701 | | | | | | | | | | | | |

TECLE: ENTRA-> CONTINUAR PF12-> IMPRIMIR PF1/6-> ALTERAR PA2 -> TERMINAR

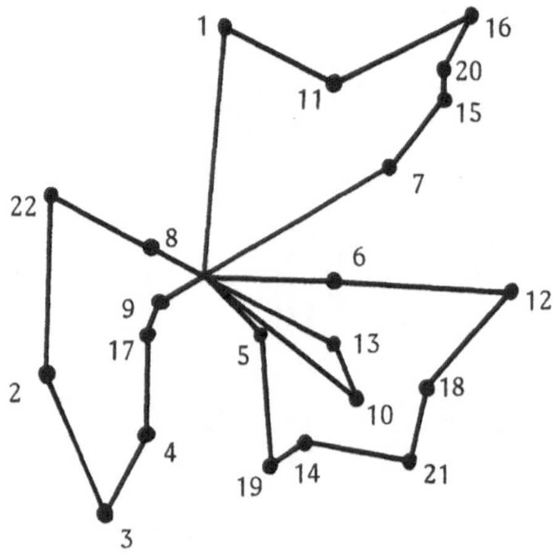


Figura 3 - Solution Routes and Their Graphical Representation

mizes the distance travelled in the new route (but without modifying the basic sequence of the route - a travelling salesman problem is *not* solved at this stage). The new set of routes is then presented to the user (Figure 4).

DATA: 24/05/84 SISTEMA INTERATIVO DE ROTEAMENTO HORA: 12:23:20

APRESENTACAO DAS ROTAS

| VEICULO | CARGA TOTAL | CAPAC. VEICULO | FOLGA CAPAC. | DIST. TOTAL | ROTA 01 | 02 | 03 | 04 | 05 | 06 | 07 | 08 | 09 | 10 | 11 | 12 |
|---|---|---|---|---|---|---|---|---|---|---|---|---|---|---|---|---|
| 1 | 1625 | 4500 | 2875 | 1783 | 1 | 11 | 16 | 20 | 15 | 7 | | | | | | |
| 2 | 2885 | 4500 | 1615 | 1645 | 6 | 12 | 18 | 21 | 14 | 19 | | | | | | |
| 3 | 3635 | 4500 | 865 | 1477 | 8 | 22 | 2 | 3 | 4 | 17 | 9 | | | | | |
| 4 | 4550 | 4500 | -50 | 766 | 5 | 10 | 13 | | | | | | | | | |
| TOTAIS: | 12695 | 18000 | 5305 | 5671 | | | | | | | | | | | | |

TECLE: ENTRA-> CONTINUAR PF12-> IMPRIMIR PF1/6-> ALTERAR PA2 -> TERMINAR
CAPACIDADE EXCEDIDA DO VEICULO

Figure 4 - Modified Solution

The new solution indicates that one of the constraints (capacity of vehicle 4) was violated. If the human scheduler judges appropriate, the solution can be adjusted through the system's facility called *demand alteration*. In this specific case the new route 4 visits clients 5, 10 and 13, with corresponding demands of 300, 4000 and 250 kg. If it is acceptable to alter the delivery of client 10 from 4000 to 3950 kg, a new feasible solution can be obtained (see Figure 5). It is important to note that with this change the total distance travelled would be reduced, and that the slack capacity of vehicle 4 would drop to zero.

```
DATA: 24/05/84           SISTEMA INTERATIVO DE ROTEAMENTO     HORA: 12:25:10
                                APRESENTACAO DAS ROTAS
        CARGA    CAPAC.   FOLGA  DIST.   _____________ R O T A _____________
VEICULO TOTAL    VEICULO  CAPAC. TOTAL   01 02 03 04 05 06 07 08 09 10 11 12
   1    1625     4500     2875   1783    1  11 16 20 15 7
   2    2885     4500     1615   1645    6  12 18 21 14 19
   3    3635     4500     865    1477    8  22 2  3  4  17 9
   4    4500     4500            766     5  10 13

TOTAIS: 12645    18000    5355   5671
TECLE: ENTRA-> CONTINUAR  PF12-> IMPRIMIR  PF1/6-> ALTERAR  PA2 -> TERMINAR
```

Figure 5 - Final Solution

## 5. CONCLUSION

Interactive computer applications represent a natural trend in the development of decision support systems, addressed to the end-user. These on-line systems are particularly useful for planning and scheduling problems of combinatorial nature, for which no exact solutions exist.

Interactive systems allow the experienced user to contribute in a positive way to improve the quality of solutions produced by the computer. The SIR system was designed to aid in the solution of vehicle routeing problems. This system should constitute a dynamic tool oriented towards the end-user, useful to route schedulers with access to a video terminal.

The specification of the system described in this paper can be useful in the definition of a similar application for use in microcomputers. This would make the software portable and available to a larger number of end-users.

REFERENCES

[ 1] BECHARA, J.J.B., Sistema interativo de roteamento - uma ferramenta de apoio à programação de entregas (Interactive Routeing System - An End-user Computer Tool for Vehicle Scheduling), M. Sc. Thesis, COPPE/UFRJ, (1984).

[ 2] BELHOT, R.V., A formação de rotas de veículos na distribuição física: modelos e metodos de solução (Vehicle Routeing in Physical Distribution: Models and Solution Methods), M. Sc. Thesis, Deptº. de Engenharia Industrial, PUC/RJ, Rio de Janeiro, (1981).

[ 3] BODIN, L., GOLDEN, B., ASSAD, A. and BALL, M., "Routing and Scheduling of Vehicles and Crews - The State of the Art", Computers and Operations Research, vol. 10, pp. 63-211, (1983).

[ 4] CLARKE, G. and WRIGHT, J. W., "Scheduling of vehicles from a central depot to a number of delivery points", Opns. Res., vol. 12, pp. 568-581, (1964).

[ 5] DANTZIG, G.B. and RAMSER, J.H., "The truck dispatching problem", Man. Sci., vol. 6, pp. 80-91, (1959).

[ 6] EILON, S., WATSON-GANDY, C.D.T. and CHRISTOFIDES, N., Distribution management: Mathematical modelling and practical analysis, London, Griffin, (1971).

[ 7] GALVÃO, R.D., "O problema do roteamento de veículos - caracterização e métodos de solução" ("The Vehicle Routeing Problem - Characterization and Solution Methods"), Proceedings of the XIII Brazilian O.R. Symposium, pp. 390-402, (1980).

[ 8] GOLDEN, B.L., MAGNANTI, T.L. and NGUYEN, H.Q., "Implementing vehicle routing algorithms", Networks, vol. 7, pp. 113-148, (1977).

[ 9] MOLE, R.H., "A survey of local delivery vehicle routing methodology", J. Opl. Res. Soc., vol. 30, pp. 245-252, (1979).

[10] STACEY, P.J., "Practical Vehicle Routeing Using Computer Programs", J. Opl. Res. Soc., vol. 34, pp. 975-981, (1983).

[11] WATERS, C.D.J., "Interactive Vehicle Routeing", J. Opl. Res.Soc., vol. 35, pp. 821-826, (1984).

[12] WATSON-GANDY, C.D.T. and FOULDS, L.R., "The vehicle scheduling problem: a survey", New Zeland Operational Research, vol. 9., pp. 73-92, (1981).

[13] WEBB, M.H.J., "Relative performance of some sequential methods of planning multiple delivery journeys", Operat. Res.Quart., vol. 23, pp. 361-372, (1972).

ON SOME ASPECTS ON SIMULATION MODELLING

Máximo Bosch P.
Depto. Ingeniería Industrial
Universidad de Chile
Santiago/Chile

The purpose of this presentation is to comment on some aspects of simulation modelling in copper foundries based on the experience in models of materials handling developed and implemented by the author in Chile. Emphasis is put on data collection procedures, and user access to the model.

1. Background

The processing of copper concentrate (from sulphide ores) into refined copper that takes place in most foundries can be represented by diagram of Fig. 1.

Copper concentrate (1) is loaded in the reverberatory furnace at a fairly constant rate. In this furnace concentrate is melted to form a liquid sulphide (matte) phase which contains all the copper of the charge; and a liquid slag phase free of copper. The slag is lighter and almost inmiscible with the matte, a property that is used to separate both phases.

Matte is normally oxidized, with air, in rotatory furnaces called converters. Converting removes the iron and the sulphur from the matte and it results in the production of blister copper (99% Cu). Converting is a batch and autogenous process, (heat is produced by the oxidation of iron and sulphur). Slag of this process is loaded back to the reverberatory furnace and blister is transferred to fire-refining furnaces, where remaining sulphur and oxygen is removed.

Materials (liquids and solids) are transferred between furnaces by bridge-cranes, in different types of containers, mainly ladles. Crane operation is an important task at the foundry because it has direct influence in production and because it is the main source of revert materials due to cooling of liquids in ladles. The latter is an important factor in measuring efficiency of the production process. Recently in most chilean foundries, a new type of furnace, (El Teniente Modified Converter) (TMC), has been implemented or is being planned to do so.

TMC is a continuous process Converter, where matte is loaded together with concentrate to produce "white metal", an intermediate product in the traditional converting process. White metal is skimmed from TMC and processed at traditional converters where blister copper is again produced.

Estimating production of a foundry in case of introduction of new technology or changes in operational rules, is a complex problem because of the large number of elements intervening and because they interralate in a random way. A digital simula-

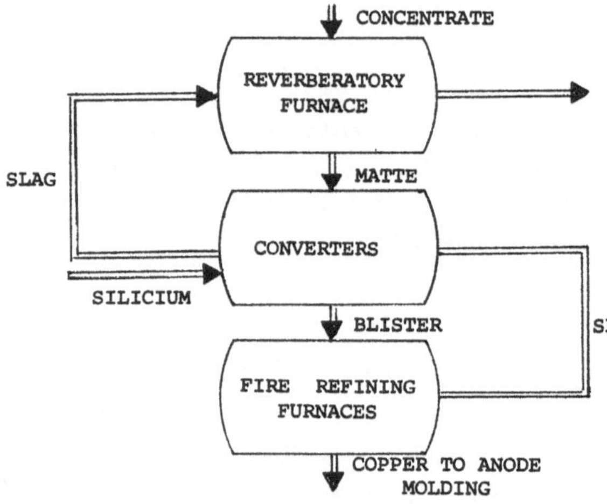


FIGURE 1 a)

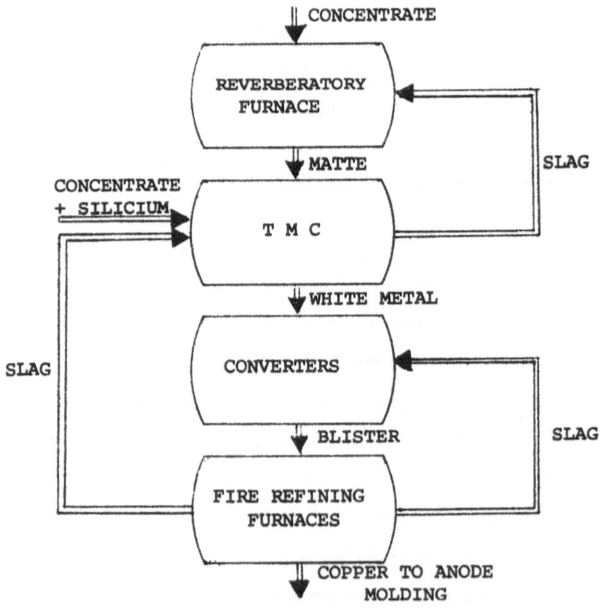


FIGURE 1 b)

tion model is clearly the tool for this problem.

## 2. The Model

The model, which is a transaction oriented, consists of a main programme and several subprogrammes. Transactions are record variables with several components, permitting, among other things: to identify the operation (transaction) by a code, to number operations with a same code, to record the time the operation goes through several procedures and, in the case of crane operations, to describe the actual location where an operation will be carried out.

The main programme basic purpose is to control time by selecting events from the future events list. Subprogrammes correspond mainly to processes that take part in the simulated elements of the foundry (reverberatory furnace, converters, modified converters, etc). Of relevance are subprogrammes that represent cranes. These subprogrammes are two, one representing the assignment of operations to the cranes and, the other, cranes activities.

Figure 2 shows a general diagram of the model. Next, we describe briefly the logic of some of the subprogrammes.

Cranes Assignment: In this subprogramme empty cranes are assigned with operations waiting in the queue, a multilevel one, in which each level corresponds to the foundry elements that demanded the operation. In this way, each operation independently of priority, will be assigned in the sequence required by the element. Operations in the first place of each level are selected using a priorities vector, a user provided vector indicating the priority of each operation in a 1 to 100 scale. However, there is also a time condition, so no operation can be pending for more than t minutes, becoming a first priority one if queueing time is larger. The highest priority operation is assigned if the "crane preference" of that operation matches the crane being assigned, where "crane preference" for every operation is defined in an array of dimensions (n x k x p), where n is the number of operations, k is the number of degrees of preferences allowed and p the number of different state conditions under which different assignment policies are possible. Operations are assigned to lower preferences cranes given certain conditions. Usually we have worked with two preference levels (k = 2).

Crane Activities: Operations assigned to bridge cranes are divided in "elementary operations" which correspond to activities carried out by the crane at a fixed location of the shed. Movement between these locations is done in a two axis system, the longitudinal representing the crane main displacement and the transversal one,

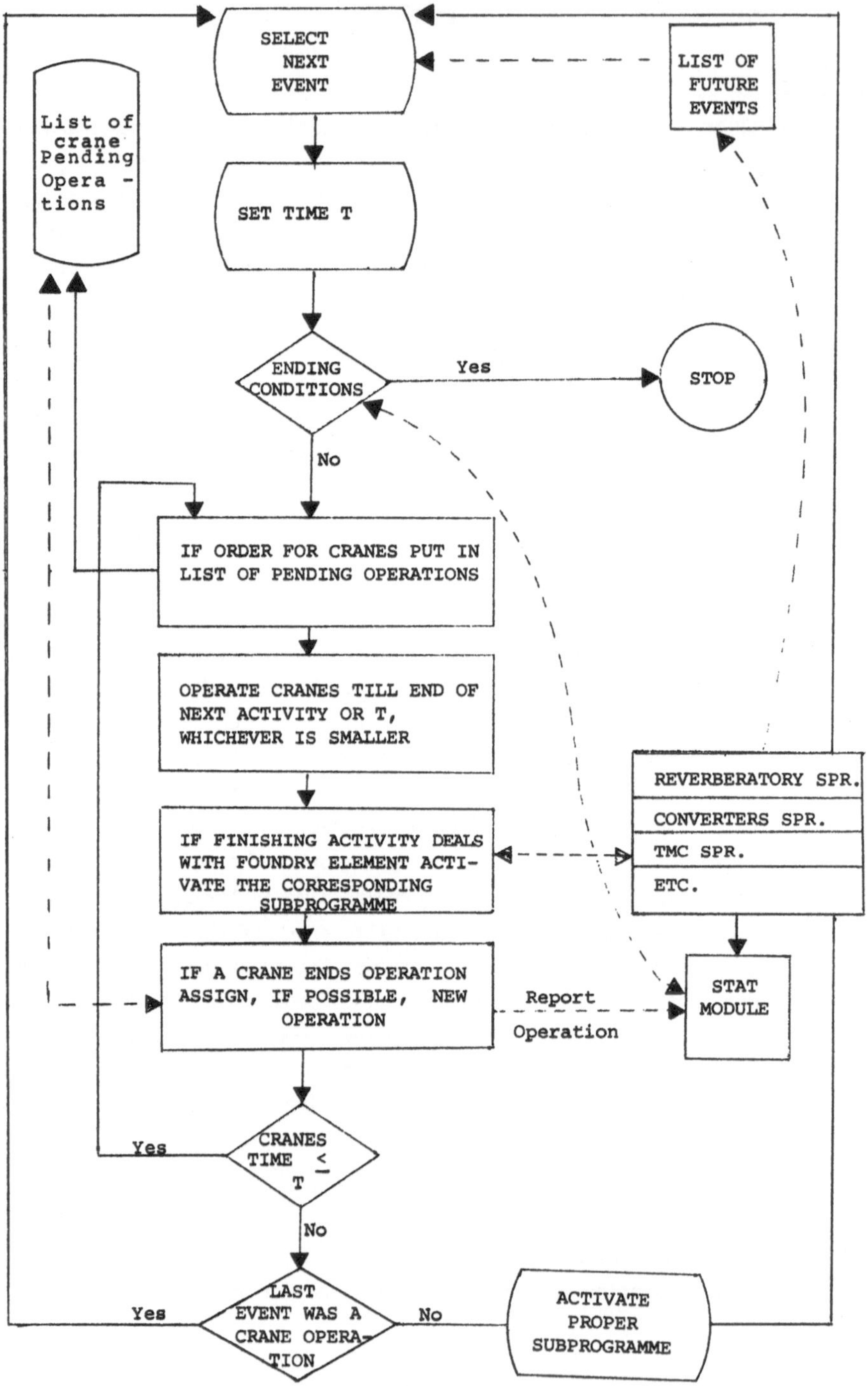
SELECT NEXT EVENT
LIST OF FUTURE EVENTS
List of crane Pending Opera - tions
SET TIME T
ENDING CONDITIONS
Yes
STOP
No
IF ORDER FOR CRANES PUT IN LIST OF PENDING OPERATIONS
OPERATE CRANES TILL END OF NEXT ACTIVITY OR T, WHICHEVER IS SMALLER
REVERBERATORY SPR.
CONVERTERS SPR.
TMC SPR.
ETC.
IF FINISHING ACTIVITY DEALS WITH FOUNDRY ELEMENT ACTI- VATE THE CORRESPONDING SUBPROGRAMME
IF A CRANE ENDS OPERATION ASSIGN, IF POSSIBLE, NEW OPERATION
Report Operation
STAT MODULE
Yes
CRANES TIME ≤ T
No
Yes
LAST EVENT WAS A CRANE OPERA- TION
No
ACTIVATE PROPER SUBPROGRAMME

FIGURE 2

the carrier's. Lifting time is included in the time of the corresponding elementa ry operation. When a bridge crane is assigned with an operation, time and location for every elementary operation is determined. Besides, for each elementary operation a priority is assigned which allows to represent and solve crane blocking. Simulated ocurrence time of each activity is recorded on the order, information that is later used together with temperature data in assessing the cooling of the liquid in the ladle, and the corresponding formation of revert materials.

Reverberatory furnace: This furnace is represented basically as an inventory of ma tte and slag. Input is due to loading of different quantities and qualities of con centrate that transforms into matte, slag and gasses.

Loading and "slagging" are managed with criteria based on stock levels of matte and slag.

Matte orders that cannot be fulfilled are stored in a queue, according to priorities provided by the user. Temperature of matte at the slagger, as well as concentrate composition are also parameters.

Converters: Converters are batch processing furnaces, and are represented by their operational cycles, which are defined by random generation of variables like effecti ve blowing time, number and type of loads and discharges. Blowing time is defined externally through parameters of a distribution function. Some output variables are: number of cycles per shift, production of blister and slag, and their copper grade. Delays are classified in crane induced delays and absence of matte (or whi te metal) delays.

TMC: Because one of the main purposes of the model is to study effective system ca pacity under different operational policies of the TMC and given the absence of his torical data, a rather detailed metallurgical model has been implemented that allows to simulate the operational policies in terms of the following variables: temperature of the furnace, volume of slag and volume of the matte-white metal bath, grade of the metal and state of traditional converters and the reverberatory furnace. De cisions that are taken include: skimming of slag and white metal, loading of concen trate and silicium, blowing (stop or re-start).

Statistical Module: In this module, transactions are processed to build the statis tics defined for output, and by the stopping rule implemented (2). The user can de fine output statistics, based in the information recorded in the order.

## 3. Data Collection Procedures.

One of the main difficulties in simulation modelling is to uncover implicit operatio nal rules. Depending on how well structured the system being simulated is implicit rules will be more or less important; however, we have found that particularly in the operation of bridge cranes they tend to be quite important. Because demands on cranes are not uniformly distributed in time, but rather come in batches, delays in processes are due to crane momentary overload. To reduce delays, rules are adapted to situational characteristics.

Because, overall, these alterations are frequent we have resorted to define situational rules, that is, a set of rules dependant of the condition of the foundry, as indicated by some relevant state variables.

Identifying these rules is part of the task of explaining some of the variance in observed processes. Analysis of data is done to find non-obvious cause-effect relationships that one might want to consider explicitly in the logic of the model. A poor analysis of the data in this sense may produce a model less sensitive to some changes than the real system, or, by unconciously including these relationships as observations in an unconditional probability distribution, later changes in the parameters of that distribution can lead to loss of relevant behaviour. We have found for example that several crane operations, specially backup operations like ladle cleaning, have a higher frequency in the first half of a shift, than in the second half. This seems to happen because there is a tendency near the end of a shift to postpone some of these operations, increasing the task load of cranes at the beginning of the next shift. If crane induced delays in converter's cycles are important this correlation may also be important.

To be able to study thoroughly the system behaviour we perform a continuous time study of the foundry. This time study is performed for about thirty consecutive shifts, by two or three time analysts per shift, to whom different elements are assigned. Time of every relevant event is recorded, with a unit measurement of one minute.

A Gantt chart of the period is built, which allows to detect errors in the data by cross checking observations of the analysts. The length of the time study is determined to have an adequate sample for most of the events of interest. For events with low frequency, additional data is required (e.g. crane maintenance), as well as for operations that take less than a minute.

The data is codified to be used by a computer program that does the Gantt chart and, also, by a general statistic package (SAS) that performs the statistical analysis.

Comparing, this Gantt chart with the one generated by the model operating under similar conditions, is an important tool in the process of model validation.

4. Accessability.

As B. Hollocks (3) has put it "...simulation is not itself a means of solving a problem, of finding the answer. It is a means of obtaining information (insight, if you wish), which in its turn contributes to the user, solving the problem". To make this true, that is to enhance "gaming" with the model as a learning process, one needs to make the model really accessible to the user.

It is our experience that in simulation models, this objective is threatened by several factors. We feel the most sensible of them are:

a) Programme changes. In many cases model modifications have to be carried out by changes in the programme itself. It is because they make this task easier that simulation languages seem to be most useful. However, the person involved in this "gaming" needs to be knowledgeable in the details of the model and the programming language.

b) Number of control variables. To test alternatives can be a tedious task, even if changes are done only through input variables, because defining a specific alternative can imply changes in a large number of control variables. For example, adding a bridge crane to an existing alternative implies to redefine input variables like priorities for solving crane's blockades, preferences in assignments and velocities of the new crane. In some alternatives one may have to redefine data for over 200 registers and, furthermore, many of these values are conditioned by the values taken by other input variables. Simulating some alternatives can become a very demanding task because the overload in input data.

c) Information required to run the model is not easily attainable. This can happen because of two reasons:

   i) Information itself is not easily available and much effort is required to obtain it, or

   ii) Information is required (or provided) by the model in unfamiliar terms. (e.g. level of matte in the reverberatory furnace is measured in the model as height, instead of tons.).

d) Low level of confidence. Statistical analysis of results is the basis for accepting or rejecting a model as valid. However, more is needed to attain user's confidence, specially when, by trying to make the model more accessible, it ends up being a "black box" from the user's standspoint. In this sense, what is needed is the capacity to verificate the model, that is, the ability to answer the question, does the model do what is supposed to do? Statistical output is not sufficient to answer this question.

To overcome the threats of the above factors we have built a support system, interface between the user and the simulator.

Our first objective in developing this system was to allow testing large number of alternatives by changing values in input variables only. However, depending on the alternative used as base case, these could mean to input from large amounts of data to just a few values. To eliminate this ambiguity and to minimize the number of changes needed to define a new alternative we have classified input variables in a five level hierarchy so that a change in a highest level variable defines, automatically, relevant values for all lower level ones.

Level one corresponds to variables that define the structure of the foundry and include variables, such as number of cranes and existing type of furnace. Each combination of values taken by these variables has an indexed list of values that are copied on the working data file.

After this input, this file contains all the information needed to run the "base alternative", including initial conditions.

Level two corresponds to variables that define minor equipment changes (in terms of the model), like cranes velocities, volume of ladles, number of each type of furnace, etc. Input at this level changes the required register on the working file.

Level three comprises operational variables. Some of them are defined by arrays (e.g. priorities of crane operations) and others by a single value (e.g. number of ladles assigned to matte transport).

Level four corresponds to parameters of probabilities distributions defining random variables. By changing a pre-specified code is possible also to redefine the type of probability distribution function.

Level five comprises variables internal to the simulation, that is, variables that regulate internal functioning of the model and do not have an equivalent in the real system, like maximum interval of time the TMC furnace can be without been accessed by the Main programme. These variables usually have meaning as such during implementation and not subsequently. Initial conditions are also defined at this level.

The simulation system is operated through a CRT terminal. The user is confronted first with having to define the base alternative he is going to operate with, that is, he has to define all level one variables. Afterwards, he can do modifications in variables of other levels. To do so he has to call the appropiate level, the current values of all variables in that level are displayed and changes are done by input of the new values in the corresponding places.

This way to operate avoids the threat posed by the first two discouraging factors mentioned above: programme changes and number of control variables.

Regarding the third factor, and though many times is not possible to get away from requiring elaborated input, we have always tried to adjust input and output variables to terms used by foundry personnel. We have found that even some straightforward transformation (e.g. meters to yards) can be a burden for someone sitting in front of the terminal. To avoid this we design an output prototype well in advance the programming stage and discuss it extensively with company engineers. Besides fulfilling this objective, this prototype has been a very useful tool in defining an agreeable final product.

Finally, as a means of verificating the model the user has the option to generate a Gantt chart covering principal elements of the foundry. By representing events graphically a full understanding of the system being tested is obtained, and, identification of bottlenecks and sources of inefficiency, a main step in developing new alternatives, it is made easier.

## 5. Conclusions.

Simulation has been found in a number of studies to be one of the most effective management science methods in use today (4). As computers, and particularly micros, are becoming more powerful, the frontiers of simulation are moving outward (3). However, larger and more complex models require stronger efforts in implementation. The measure of successful implementation is the degree to which the manager's problem solving is augmented and stimulated rather than wether the conclusion's of the management scientist are adopted (5). Thus, continuing use (6) of the model is a key objective in implementation. Based on the experience gained in developing models for a particular problem, we have described steps that we regard as pivotal in explaining successful implementation.

## REFERENCES.

1) A.K. Biswas y W.G. Davenport. Extractive Metallurgy of Copper. 2nd Edition. Pergamon Press 1980.
2) T.H. Naylor (1971). Computer Simulation Experiments with Models of Economic Systems. pp. 293-294. Wiley, New York.
3) B.W. Hollocks (1984). Simulation and the Micro. Journal of the Operational Research Society. Vol. 34, N° 4.
4) D.P. Christy, H.J. Watson (1983). The Application of Simulation: A Survey of Industry Practice. Interfaces. Vol. 13, N° 5.

5) J.S. Hammond (1974). The Roles of the Manager and Management Scientist in Successful Implementation. Sloan Management Review.

6) G.L. Urban (1974). Building Models for Decision Makers. Interfaces, Vol. 4, N° 3.

# SIMULATION OF MONTHLY FLOW SERIES USING PRECIPITATION INPUTS GENERATED BY ARMA MODELS

E. Brown and X. Vargas
Centro de Recursos Hidráulicos
Departamento de Ingeniería Civil
Universidad de Chile
Casilla 5373 Santiago Chile

## Introducción.

One of the frequently encountered problems in applied hydrology is to provide adequate design time series of flows in fixed cross sections of rivers, to test and or simulate the operation of different alternative designs for certain hydraulic structures such as reservoirs and hydroelectric plants. For many of these situations, the mean monthly flow series are currently used. A common practice consists of generating synthetic series of monthly flows, which intend to preserve some of the statistical properties derived from the historic series available for the point of interest. Usually the parameters which are retained for each monthly time period are the mean and standard deviation; some models consider additionally the skewness coefficient, autocorrelation coefficients and a desired probability density function.

The problem of generating synthetic flow series has been approached in different ways; probably the best known and most widely used approach is the Markov lag-one model (Fiering & Jackson, 1971). This model however has some drawbacks which are discussed, among others, by Askew et al (1970) and by Brown and Torretti (1977). Other authors have approached the problem by generating precipitation series and using a rainfall-runoff relation or model, to produce the flow sequences (Ott, 1971; Arrese, 1984); the models used for the generation of the rainfall series have included Markov chain models or similar, models based on a Poisson process, etc.

In general in these approaches no analysis has been done so as to indentify an "adequate" model for the precipitation time series under study; rather, a predetermined hypothesis has been made to that effect.

In this paper a procedure for generating monthly streamflow series, through the combined use of monthly precipitarion series generared by ARMA models, and a rainfall-runoff model, is presented. The ARMA model that is used in each case, is that which can be identified from the original precipitation time series.

Modelling steps.

A nonseasonal ARMA (p,q) model ban be expressed as

$$\phi(B)\tilde{Z}_t = \theta(B)a_t \tag{1}$$

in which $\phi(B)$ and $\theta(B)$ are the autoregresive (AR) polynomial of order p, and the moving average (MA) polynomial of order q respectively; B is the backward shift operator defined by $B^i\tilde{Z}_t = \tilde{Z}_{t-i}$; $\tilde{Z}_t = Z_t - \bar{Z}$ where $Z_t$ is the time series to be modelled and $\bar{Z}$ is its mean; $a_t$ is a normally distributed, independent white noise residual with mean 0 and variance $\sigma_a^2$.

The first step in the modelling, is the determination of the order of the polynomials that adequately represent the time series under study This is done through the analysis of four functions (Hipel et al, 1977) autocorrelation (ACF); partial autocorrelation (PACF), inverse autocorrelation (IACF) and inverse partial autocorrelation (IPACF). Once identified, and its parameters calculated by minimizing the unconditional sum of squares function, through the use of the Marquardt algorithm, the model should be verified through the analysis of the ACF of the residuals, considering that these must be independent, homocedastic and normally distributed.

The conditions of normal distribution of the residuals, can be achieved by and appropriate transformation of the basic time serie ($X_t$). If a transformation is needed, when in the process of generating synthetic time series, it is of interest to preserve the statistical properties of the original series ($X_t$). In particular for hydrologic series, where these properties are significantly different from one monthly period ($\tau$) to another, at least two of the following conditions should be imposed to the generated series ($W_t$), for each t corresponding to month $\tau$:

$$E(W_t) = \bar{X}_\tau \tag{2}$$

$$E[(W_t - \bar{X}_\tau)^2] = s_{x\tau}^2 \tag{3}$$

$$\frac{E[(W_t - \bar{X}_\tau)^3]}{[E[(W_t - \bar{X}\ )^2]]^{3/2}} = \gamma_{x\tau} \tag{4}$$

in which $\bar{X}_\tau$, $s^2_{X\tau}$, $\gamma_{X\tau}$, are the mean, variance and skewness coefficient respectively, corresponding to month $\tau$ in the original serie.

For hydrologic series the logarithmic and square root transformations are commonly used. For these, the imposed conditions (2), (3) and (4), for each time t corresponding to a particular month $\tau$, lead to:

a) Transformation $Y_t = \text{Ln}(X_t + c_\tau)$

$$\bar{X}_\tau = \exp\,(\bar{Y}_\tau + s^2_{y\tau}/2) - c_\tau \tag{5}$$

$$s^2_{X\tau} = \exp\,(2\bar{Y}_\tau + s^2_{y\tau})[\exp(s^2_{y\tau}) - 1] \tag{6}$$

$$\gamma_{X\tau} = \frac{\exp(3s^2_{y\tau}) - 3\exp(s^2_{y\tau}) + 2}{[\exp(s^2_{y\tau}) - 1]^{3/2}} \tag{7}$$

b) Transformation $Y_t = \sqrt{X_t} + c_\tau$

$$\bar{X}_\tau = s^2_{y\tau} + \bar{Y}^2_\tau - c_\tau \tag{8}$$

$$s^2_{X\tau} = 2s^2_{y\tau}(s^2_{y\tau} + 2\bar{Y}^2_\tau) \tag{9}$$

$$\gamma_{X\tau} = \frac{8s_{y\tau}(s^2_{y\tau}+3\bar{Y}^2_\tau)}{[2(s^2_{y\tau} + 2\bar{Y}^2_\tau)]^{3/2}} \tag{10}$$

It must be emphasized that if the lower bound parameter $c_\tau$ is made equal to zero, then, only the first two moments are preserved.

In the formulation of the model it is convenient to work with standardized series:

$$Z_t = \frac{Y_t - \bar{Y}_\tau}{s_{y\tau}} \tag{11}$$

Hence, for the generation process, $Z_t$ must be a normally distributed independent variable with mean zero and unit variance, so as to preserve $\bar{Y}_\tau$ and $s_{y\tau}$. To achieve this, the residuals ($a_t$) in the model, should be generated with an appropriate value of their variance ($\sigma^2_a$). This value should be:

$$\text{(i)}\quad \sigma^2_a = (1 - \rho_1\phi_1 - \rho_2\phi_2 - \ldots - \rho_p\phi_p) \tag{12}$$

for pure AR models, where $\rho_i$ is the correlation coefficient of lag i.

$$\text{(ii)}\quad \sigma^2_a = (1 + \theta^2_1 + \theta^2_2 + \ldots + \theta^2_q)^{-1} \tag{13}$$

for pure MA models.

In the case of ARMA models the variance must be obtained from the equation:

$$\gamma_0 = \sigma_z^2 = \phi_1\gamma_1 + \dots + \phi_p\gamma_p + \sigma_a^2 - \theta_1\gamma_{za}(-1) - \dots - \theta_q\gamma_{za}(-q) \quad (14)$$

in which the autocovariance function is defined by:

$$\gamma_k = \phi_1\gamma_{k-1} + \dots + \phi_p\gamma_{k-p} + \gamma_{za}(k) - \theta_1\gamma_{za}(k-1) - \\ - \dots - \theta_q\gamma_{za}(k-q) \quad (15)$$

in which $\gamma_{za}(k) = E[\tilde{Z}_{t-k}a_t]$ is the cross covariance function between $\tilde{Z}$ and a.

Case Studies and Results.

The procedure described above has been applied to the generation of streamflow sequences for two pluvial catchments located in the Central Zone of Chile: Purapel and Porvenir. For these, the rainfall stations of Constitución and Loncha were used respectively, to formulate the appropriate ARMA models for precipitation inputs to the catchments. Once the ARMA model was identified in each case and its parameters calculated, 50 series of 50 years each were generated for each catchment; in fact 700 monthly values were generated each time and the first 100 were disregarded for the analysis. A rainfall-runoff model based on that presented by Ferrer et al (1973) was used to turn precipitation into monthly streamflow.

A preliminary analysis of the basic precipitation series determined the need of transforming them, in order to achieve normality of the residuals. In Table 1 the models identified for each station are given, together with the values of their parameters and transformations used. In all cases the Portemanteau test for independence of the residuals was satisfied.

In the case of Constitución station the standard identification routine indicated an MA (10) model. However, as this was considered not very appealing from a physical point of view, an MA (5) model wich appeared reasonable, was also fitted.

In the generation procedure, specially in those time period (months) with very low precipitation, some negative values occurred. As this is not physically possible, a standard lower limit of zero was adopted in these cases, for the final value of monthly precipitation. For this reason, the values of the mean and standard deviation of the generated precipitation series are slightly higher and lower respectively, than that of the basic historic record. However, the differences in this aspect can be considered insignificant; this is specially true when the basic historic record is such that is does not include months with zero (or nearly) mean values.

TABLE 1

CHARACTERISTICS OF THE IDENTIFIED MODELS

| STATION | MODEL | PARAMETERS | TRANSFORMATION |
|---|---|---|---|
| Loncha | MA(4) | $\theta_1=0.0$; $\theta_2=-0.121$; $\theta_3=-0.083$; $\theta_4=-0.133$ | Logarithmic |
| Constitución | MA(10) | $\theta_1=-0.160$, $\theta_2=-0.200$; $\theta_3=0.3$ $\theta_4=0.126$, $\theta_5=-0.159$; $\theta_6=\theta_7=0.0$; $\theta_8=-0.190$; $\theta_9=0.0$; $\theta_{10}=0.265$ | Square Root |
| | MA(5) | $\theta_1=-0.140$, $\theta_2=-0.122$; $\theta_3=\theta_4=0.0$; $\theta_5=-0.135$ | Square Root |

In Table 2 some of the results at the annual level, for the generated streamflow series are shown for each of the two sites. In the case of Purapel, results corresponding to the two ARMA models that were identified and also results obtained by Arrese (1984) using a Markov chain model for daily precipitation are included. The statistical properties that are compared in each line of Table 2 include the historic mean ($\bar{q}$) and the average generated mean ($\bar{\bar{w}}_q$) in line 1, standard deviation of the mean values generated ($s_{\bar{w}q}$) in line 2; historic standard deviation ($s_q$) and the mean generated standard deviation ($\bar{s}_{wq}$) in line 4; standard deviation of the generated standard deviations ($s_{swq}$) in line 5; seasonal correlation coefficient ($r_{iv}$) calculated between the winter mean flow and the summer mean flow within the same hydrologic year, historic and mean generated value, in line 7; and, standard deviation of the generated seasonal correlation coefficient ($s_r$) in line 8. In all cases, the historic values were obtained through the rainfall-runoff simulation model, using the historic precipitation series.

TABLE 2

STREAMFLOW ANNUAL RESULTS

| LINE N° | STATISTIC | PURAPEL AT NIRIVILO (mm) | | | | PORVENIR I DAM CATCHMENT ($m^3/s$) | |
|---|---|---|---|---|---|---|---|
| | | Hist. | MA(10) | MA(5) | Markov | Hist. | MA(4) |
| 1 | Mean | 20.4 | 20.7 | 20.7 | 19.6 | 0.83 | 0.88 |
| 2 | $s_{\bar{w}q}$ | - | 1.24 | 1.37 | 2.30 | - | 0.10 |
| 3 | $\frac{\bar{q}-\bar{\bar{w}}_q}{\bar{q}}$ (%) | - | -1.5 | -1.5 | 3.9 | - | -6.0 |
| 4 | Stand. Dev. | 11.0 | 10.2 | 10.5 | 9.0 | 0.71 | 0.69 |
| 5 | $s_{swq}$ | - | 1.15 | 1.22 | 1.3 | - | 0.10 |
| 6 | $\frac{s_q-\bar{s}_{wq}}{s_q}$ | - | 7.0 | 4.6 | 18.2 | - | 2.8 |
| 7 | $r_{iv}$ | 0.614 | 0.602 | 0.633 | - | 0.723 | 0.648 |
| 8 | $s_r$ | - | 0.084 | 0.077 | - | - | 0.079 |

Analyzing Table 2, it is apparent that in terms of annual values, results compare favorably with the historic values. This is specially true in the case of Purapel which corresponds to a more humid area. Also, the values generated using the ARMA models for monthly precipitarion series, behave better than those which were generated with a Markov chain model for daily precipitation. The results which use the two different ARMA models for Purapel (runs were made using exactly the same random seed for each of the 50 series) differ very slightly between them. Finally, the generated values retain adequately the seasonal correlation coefficient, fact which is appealing for an applied hydrologist.

In figures 1 to 4 a sample of the results at the monthly level is shown. In figure 1, runoff at Purapel at Nirivilo station for each monthly time period is plotted, including results by Arrese (1984) and historic values. In figure 2 percentual differences between the simulated flows with the historic records and average simulated flows using the generated precipitation series are shown, for both sites. In figure 3 an analogous comparison is made for standard deviations. Finally, in figure 4 the average lag-one correlation coefficient, for the monthly flows generated for Purapel at Nirivilo are presented.

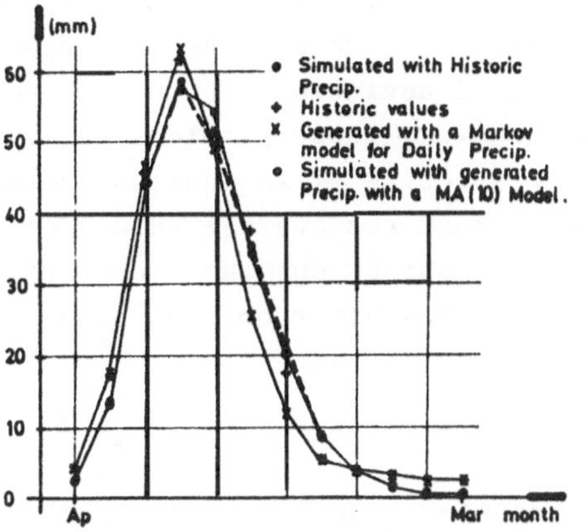


Fig. 1 PURAPEL AT NIRIVILO STATION AVERAGE MEAN MONTHLY FLOWS

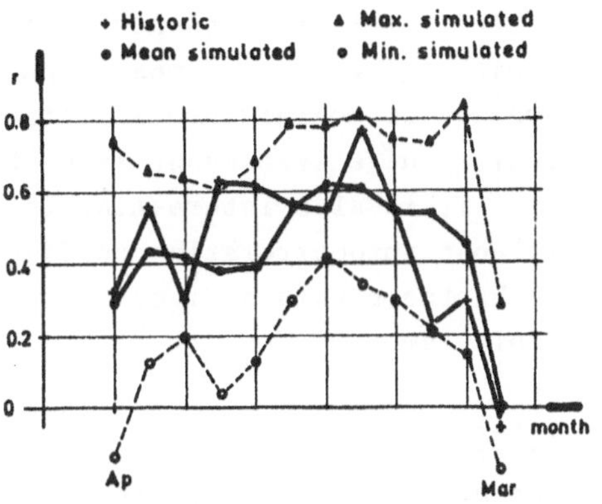


Fig. 4 PURAPEL AT NIRIVILO. AVERAGE MONTHLY LAG ONE CORRELATION COEFFICIENT

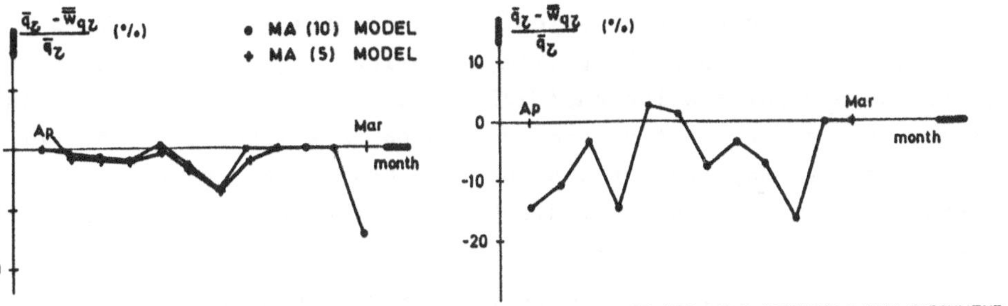


a) PURAPEL AT NIRIVILO

b) RUNOFF FROM THE PORVENIR I DAM CATCHMENT

Fig. 2 MONTHLY PERCENTUAL ERROR OF THE SIMULATED MEAN FLOWS

$\bar{q}_\tau$ : mean monthly simulated with historic precipitation

$\bar{\bar{w}}_{q\tau}$: average mean monthly flow simulated with generated precipitation.

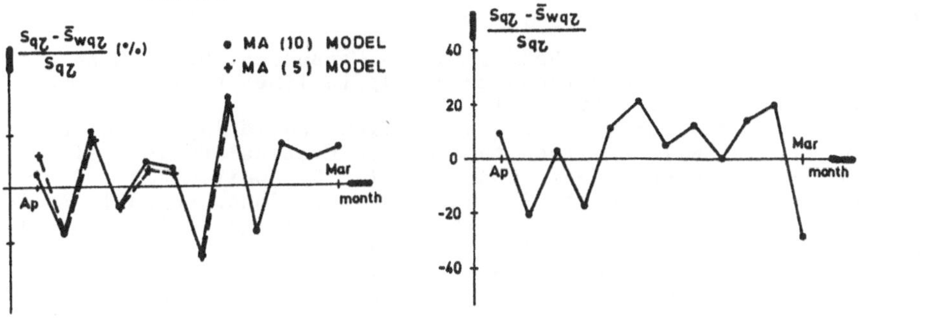


a) PURAPEL AT NIRIVILO

b) RUNOFF FROM THE PORVENIR I DAM CATCHMENT

Fig. 3 MONTHLY PERCENTUAL ERROR OF THE STANDARD DEVIATION

$s_{q\tau}$ : standard deviation of historic mean monthly flows

$\bar{s}_{wq\tau}$: mean standard deviation of the generated mean monthly flows

The analysis of the figures shows that results in terms of monthly values, also retain the essence of the historic record. The main problems with the generated means and standard deviations, occur for near zero values of monthly precipitation; however, these problems can be considered quite irrelevant from the perspective of an applied hydrologist. It is also interesting to note, that the average values of the lag-one autocorrelation coefficient, smoothes out the fluctuations of the historic values, fact that seems reasonable from a physical point of view.

## Final Remarks.

Judging by the limited results here presented, the use of appropriately identified and adequately fitted ARMA models for monthly precipitation, together with a rainfall-runoff model, can produce synthetic monthly streamflow series which retain with reasonable precision the values of the principal statistics of the historic record. It is of special interest and appealing from a physical standpoint, that seasonal and lag-one correlation coefficients are preserved properly. In this respect, the generating procedure used here seems to be better than alternative ways to achieve the same objectives; particularly it surmounts one of the important drawbacks of the Markov lag-one model for direct streamflow synthesis.

In very arid zones, or at sites where very dry seasons occur, the generated series can deliver a significant percentage of negative values, which have to be dealt with in a conventional way, like the one employed here, wich can produce undesirable alterations in overall results.

## References.


- Arrese, J.: Generación de Gastos Medios Mensuales. Simulación Hidrológica con Series Estocásticas de Precipitaciones Diarias. Memoria de Ing. Civ. (1984). Univ. Chile.
- Askew, A.; W. Yeh and W. Hall.: Streamflow Generating Techniques: A Comparison of their Abilities to Simulate Critical Periods of Drought. Water Resour. Cent. Cont. 131 (1970). Univ. Calif.
- Brown, E., and E. Torretti: Algunas Consideraciones sobre la Aplicación del Modelo de Markov en Generación Estocástica de Caudales. III Col. Nac. SOCHID. Stgo., Chile (1977) 375.
- Ferrer, P.; E. Brown and L. Ayala: Simulación de Gastos Medios Mensuales en una Cuenca Pluvial. II Col. Nac. SOCHID. Stgo., Chile. 2 (1973) 3.49.
- Fiering, M. and B. Jackson: Synthetic Streamflow. Am. Geophys. Union. Water Resour. Monogr. 1 (1971).
- Hipel K., A. Mc Leod and W. Lennox: Advances in Box-Jenkins Modelling 1: Model Construction. Water Resour. Res. 13 (1977) 567.
- Ott. R.: Streamflow Frecuency Using Stochastically Generated Hourly Rainfall. Stanford Univ. Tech. Rep. 151 (1971).

# ON THE BOUNDEDNESS OF CERTAIN POINT-TO-SET MAPS AND ITS APPLICATION IN OPTIMIZATION.

Luis Contese
Departamento de Matemáticas y Ciencias de la Computación
Facultad de Ciencias Físicas y Matemáticas
Universidad de Chile
Casilla 170 - Correo 3, Santiago de Chile

SUMMARY.

A general result concerning the local uniform boundedness of the set of approximate optimal solutions of a continuously parametrized family of optimization problems is given. This result extends Hogan's Theorem and contains, in particular, the already classical local and global boundedness results for the level sets of the quasiconvex and convex functions, respectively. Also, a result concerning the existence of minimizing points for partially inf-bounded functions is obtained as a straightforward consequence. Finally, some general stability properties for a continuously parametrized family of convex optimization problems and their corresponding duals are obtained.

INTRODUCTION.

Given a parametrized family of maximization problems of the form $P(y)$: Max $\{f(x,y)/x \in \Omega(y)\}$, where X and Y are any two metric spaces, $f: X \times Y \to \overline{\mathbb{R}}$ an extended numerical function and $\Omega : Y \longrightarrow X$ some point-to-set map, we are often interested in studying, at first, the continuity properties of the marginal value function $v(y) =: \sup\{f(x,y)/x \in \Omega(y)\}$ under certain continuity hypothesis on $f(\cdot,\cdot)$ and $\Omega(\cdot)$. Indeed, the most elementary stability properties of the family $P(y)$ are naturally defined in terms of the continuity properties of this function. In this respect, it is well known (see [1],[8],[9],[2] ) that if $f(\cdot,\cdot)$ is lower-semicontinuous over $\Omega(\bar{y}) \times \{\bar{y}\}$ and if $\Omega(\cdot)$ is lower-continuous at $\bar{y}$, then $v(\cdot)$ is lower-semicontinuous at $\bar{y}$. On the contrary, if $f(\cdot,\cdot)$ is upper-semicontinuous and if $\Omega(\cdot)$ is upper-continuous (or, closed) at $\bar{y}$, we need some additional hypothesis, as the local uniform boundedness of $\Omega(\cdot)$ at $\bar{y}$, in order to assert the upper-semicontinuity of $v(\cdot)$ at $\bar{y}$. With this idea in mind, we give in Theorem 1 a general

Partially supported by Fondo Nacional de Ciencias under Grant Nº 1142 and Departamento de Investigación y Bibliotecas under Grant E2263-8623.

result concerning the local uniform boundedness of the $\varepsilon$-optimal solutions point-to-set map ar some $\bar{y}$ Y. In the case of quasi-concave economic functions, this Theorem generalizes Hogan's Theorem [6,Theorem 9] with still weaker hypothesis than those given in [2, Theorem 10].

On the other hand, in the case of concave economic functions, Theorem 1 gives a global boundedness result in terms of the approximation threshold. As a consequence of this, the classical local and global boundedness results for the level sets of a quasiconcave and concave function respectively ([4],[5]) are simultaneously recovered. Moreover, in Corollary 1.1, the local uniform boundedness of a continuous convex valued point-to-set map is characterized, while, in Corollary 1.2, the local inf-boundedness of a continuous parametrized convex function is stated. Next, in Theorem 2, an application of Theorem 1 to the proof of existence of minimizing points for partially inf-bounded functions is given. An example stressing the importance of the quasiconvexity hypothesis in this last Theorem is also given. Previous to proving Theorem 1, we state several Lemmas that point out different ideas that seem to us interesting in their own. This is particularly the case of Lemma 3, stating the continuity of the recession cone map of a continuously parametrized family of convex sets and of Lemma 4, characterizing the compactness of the approximate optimal solution set for an homogeneous concave function over a closed convex set.

In Theorem 3, we study the continuity of the set of feasible solutions of a continuously parametrized family of inequality systems when the variables are also constrained to belong to some continuously parametrized closed set. This result is a straightforward generalization of the classical result where this set is a closed constant one. An application of Theorem 3 is found in Theorem 4, where a general stability result for a continuously parametrized family of convex problems is given. Concretely, Lemma 6 states that if, for some value $\bar{y}$ of the parameter y, the constraints of problem $P(\bar{y})$ verify the Slater Constraint Qualification, the same is true for all neighboring problems $P(y)$, with y near $\bar{y}$. Under these circumstances, Theorem 4 asserts that the marginal value function of the family $P(y)$ is continuous at $\bar{y}$. Furthermore, it asserts that its values coincide with the values of the corresponding dual marginal function, on some neighborhood of $\bar{y}$ or, equivalently, that there is no duality gap on some neighborhood of $\bar{y}$. In this sense, Theorem 4, as Theorem 10 [2], generalizes the original result given by Hogan [7, Lemma 2] for the constant closed convex set constrained parametrized inequality systems, to the continuously parametrized convex set case.

Notation, Definitions and Review.

Given two arbitrary metric spaces X and Y, a point-to-set map $\Omega: Y \rightarrow X$ is said to be:

i) upper-continuous or closed at $\bar{y} \in Y$, if for any sequence $\{y^k \in Y\}_{k \in N}$ converging to $\bar{y}$ and for any sequence $\{x^k \in \Omega(y^k)\}_{k \in N}$ converging to $\bar{x}$, we have $\bar{x} \in \Omega(\bar{y})$,

ii) lower-continuous at $\bar{y} \in Y$, if for any sequence $\{y^k \in Y\}_{k \in N}$ converging to $\bar{y}$ and for any $\bar{x} \in \Omega(\bar{y})$, there exists a sequence $\{x^k \in X\}_{k \in N}$ converging to $\bar{x}$ and some $\bar{k} \in \mathbb{N}$ such that $x^k \in \Omega(y^k)$ for all $k \in \mathbb{N}$, $k \geq \bar{k}$,

iii) lower-semicontinuous at $(\bar{y}, B)$, where $\bar{y} \in Y$, $B \subset X$, if for any sequence $\{y^k \in Y\}_{k \in N}$ converging to $\bar{y}$, there exists $\bar{x} \in B$, a subsequence $\{y^k \in Y\}_{k \in N'}$ and a sequence $\{x^k \in X\}_{k \in N'}$ converging to $\bar{x}$ such that $x^k \in \Omega(y^k)$, $\forall k \in N'$,

iv) locally uniformly bounded [resp. compact] at $\bar{y}$, if there exists a neighborhood $N(\bar{y})$ of $\bar{y}$ and a bounded [resp. compact] subset K of X such that $\Omega(y) \subset K$, $\forall y \in N(\bar{y})$.

Next, given the parametrized family of optimization problems:
$P(y)$: Max $\{f(x,y) / x \in \Omega(y)\}$, where $f: X \times Y \rightarrow \bar{\mathbb{R}}$ is an extended real valued function, the supremal value function $v: Y \rightarrow \bar{\mathbb{R}}$ associated with this family is defined as:

$$v(y) = \begin{cases} \sup\{f(x,y) / x \in \Omega(y)\}, & \text{if } \Omega(y) \neq \phi \\ -\infty, & \text{if } \Omega(y) = \phi \end{cases}$$

and the quasi-optimal solutions map $M: Y \times \mathbb{R}_+ \rightarrow X$, where $\mathbb{R}_+ =: \{\rho \in \mathbb{R} / \rho \geq 0\}$, as the point-to-set map giving, for any $(y, \varepsilon) \in Y \times \mathbb{R}_+$, the set of all the $\varepsilon$-optimal solutions for problem $P(y)$ i.e. $M(y,\varepsilon) =: \{x \in \Omega(y) / v(y) \leq f(x,y) + \varepsilon\}$. Accordingly, the optimal solutions map $M: Y \rightarrow X$, giving for any $y \in Y$, the set of all the optimal solutions for $P(y)$, is defined by $M(y) =: M(y,0)$, $\forall y \in Y$.

We remind now some classical results concerning the continuity properties of the supremal value function as well as of the quasi-optimal solutions map in terms of the continuity properties of $f(\cdot,\cdot)$ and $\Omega(\cdot)$. These results are straightforward variants of Berge Maximum Theorems [ 1,8 ] as those considered in [2 ]. An exhaustive study of the continuity of the supremal value function is given by J.P. Penot [9 ].

Theorem A1: Let $f(\cdot,\cdot)$ and $\Omega(\cdot)$ be as defined above. Then, if

i) $f(\cdot,\cdot)$ is lower-semicontinuous over $\Omega(\bar{y}) \times \{\bar{y}\}$ and if

ii) $\Omega(\cdot)$ is lower-continuous at $\bar{y}$ or, alternatively, if

ii)' $\forall\ \varepsilon > 0$, $\exists\ \bar{\varepsilon} \in [0,\varepsilon]$ such that $\Omega(\cdot)$ is lower-semicontinuous at $(\bar{y},M(\bar{y},\bar{\varepsilon}))$, we have that $v(\cdot)$ is lower-semicontinuous at $\bar{y}$.

Proof: From the definition of $v(\cdot)$, if $\Omega(\bar{y}) = \phi$, then $v(\bar{y}) = -\infty$ and the result trivially follows in this case. Let us then suppose that $\Omega(\bar{y}) \neq \phi$. Now, let $\{y^k \in Y/k\in N\}$ be a sequence converging to $\bar{y}$ such that the sequence $\{v(y^k)/k\in N\}$ is also convergent in $\bar{\mathbb{R}}$. Under the hypothesis ii)', for any $\varepsilon > 0$ there exists $\bar{\varepsilon} \in [0,\varepsilon]$, $\bar{x} \in M(\bar{y},\bar{\varepsilon})$, a subsequence $\{y^k \in Y/k \in N'\}$ and a sequence $\{x^k \in X/k \in N'\}$ converging to $\bar{x}$ such that $x^k \in \Omega(y^k)$, $\forall k\in N'$. Then, from the lower-semicontinuity of $f(\cdot,\cdot)$ at $(\bar{x},\bar{y})$ we have $v(\bar{y}) \leq f(\bar{x},\bar{y})+\bar{\varepsilon} \leq \liminf_{k\to\infty,k\in N'} f(x^k,y^k)+\bar{\varepsilon} \leq \liminf_{k\to+\infty,k\in N'} v(y^k)+\bar{\varepsilon} =$

$\lim_{k\to+\infty,k\in N} v(y^k)+\bar{\varepsilon}$, for some $\bar{\varepsilon} \in [0,\varepsilon]$. But since this is true for any $\varepsilon > 0$, we necessarly have $v(\bar{y}) \leq \lim_{k\to+\infty,k\in N} v(y^k)$. We have then proved the result under the hypothesis i) and ii)' and, in particular, under the hypothesis i) and ii) in the case where $v(\bar{y})<+\infty$. In fact, in this case, the hypothesis ii) implies the hypothesis ii)' (Why?). Let us finally prove the result under the hypothesis i) and ii), for the case $v(\bar{y})=+\infty$. In this situation, there exists a sequence $\{x^j \in \Omega(\bar{y})/j \in J\}$ such that $\{f(x^j,\bar{y})/j \in J\}$ converges to $+\infty$. Now, since $\Omega(\cdot)$ is lower-continuous at $\bar{y}$, we may choose, for each $j \in J$, a sequence $\{x^{j,k}\in\Omega(y^k)/k\in N\}$ converging to $x^j$ when $k$ converges to $+\infty$. Then, from the lower-semicontinuity of $f(\cdot,\cdot)$ at $(x^j,\bar{y})$ we have that $f(x^j,\bar{y}) \leq \liminf_{k\to+\infty} f(x^{j,k},y^k)$ $\leq \liminf_{k\to+\infty} v(y^k)$ for any $j \in J$, so that $\liminf_{k\to+\infty} v(y^k) \geq +\infty$, from where the result follows.

Remark. We may easily verify that in the case where $v(\bar{y})<+\infty$, the hypothesis ii) in Theoren A1 implies the alternative hypothesis ii)'.

Theorem A2: Let $f(\cdot,\cdot)$ and $\Omega(\cdot)$ be as previously defined. Then, if

i) $f(\cdot,\cdot)$ is upper-semicontinuous over $\Omega(\bar{y}) \times \{\bar{y}\}$,

ii) $\Omega(\cdot)$ is upper-continuous at $\bar{y}$, and if,

iii) $\Omega(\cdot)$ is locally uniformly compact at $\bar{y}$ or, alternatively, if

iii)' there exists a compact subset $K$ of $X$ and a neighborhood $N(\bar{y})$ of $\bar{y}$ in $Y$ such that $M(y)\neq\phi$ and $M(y)\subset K$, $\forall y\in N(\bar{y})$, we have that $v(\cdot)$ is upper-semicontinuous at $\bar{y}$.

Proof: See [1,2] for example.

Theorem A3: Let $f(\cdot,\cdot)$ and $\Omega(\cdot)$ verify: i) $f(\cdot,\cdot)$ is continuous over $\Omega(\bar{y})x\{\bar{y}\}$; ii) $\Omega(\cdot)$ lower-continuous at $\bar{y}$ or, alternatively, if $\forall\varepsilon>0$, $\exists\bar{\varepsilon}\in[0,\varepsilon]$: $\Omega(\cdot)$ is lower-semicontinuous at $(\bar{y},M(\bar{y},\bar{\varepsilon}))$, and iii) $\Omega(\cdot)$ is upper-continuous at $\bar{y}$. Then, $M: Y x \mathbb{R}_+ \to X$ is upper-continuous over $\{\bar{y}\} x \mathbb{R}_+$.

Proof: The result is a straightforward consequence of Theorem A1.

Lemma 1: Let W and Y be arbitrary metric spaces and let the relations $\Omega: Y \rightarrow\!\!\triangleright W$, $f: W \times Y \to \bar{\mathbb{R}}$ verify, at a given point $\bar{y} \in Y$ :

i) $\Omega(\bar{y})$ is a closed subset of W,
ii) $f(\cdot,\bar{y})$ is upper-semicontinuous over $\Omega(\bar{y})$.

Then, the partial point-to-set map $M(\bar{y},\cdot)$, defined over $\mathbb{R}_+$, is such that:

a) $M(\bar{y},\cdot)$ is closed valued and upper-continuous over $\mathbb{R}_+$,

b) $M(\bar{y},\varepsilon_o)$ is nonempty and compact for some $\varepsilon_o > 0$ if and only if $M(\bar{y},\varepsilon)$ is nonempty and compact for all $\varepsilon \in [0,\varepsilon_o]$.

Proof: a) The closedness of $M(\bar{y},\varepsilon)$, for all $\varepsilon \in \mathbb{R}_+$, is a straightforward consequence of the hypothesis i) and ii). Now, let $\varepsilon \in \mathbb{R}_+$ and let $\{\varepsilon_k \in \mathbb{R}_+\}_N$, $\{w^k \in W\}_N$ be sequences such that $\varepsilon_k \to \varepsilon$, $w^k \to \bar{w}$ when $k \to +\infty$, with $w^k \in M(\bar{y},\varepsilon_k) \subset \Omega(\bar{y})$, $\forall k \in N$. From hypothesis i) we then have $\bar{w} \in \Omega(\bar{y})$. Moreover, since $v(\bar{y}) \leq f(w^k,\bar{y}) + \varepsilon_k$, $\forall k \in N$, from hypothesis ii) we have that $v(\bar{y}) \leq f(\bar{w},\bar{y}) + \varepsilon$. Consequently, $\bar{w} \in M(\bar{y},\varepsilon)$ as was to be proved.

b) If $v(\bar{y}) = +\infty$, the result is trivially true. If $v(\bar{y}) < +\infty$, there exists a sequence $\{\varepsilon_k\}_N$, $\varepsilon_k \in ]0,\varepsilon_o]$, $\forall k \in N$, and a sequence $\{w^k \in W\}_N$ such that $w^k \in M(\bar{y},\varepsilon_k)$, $\forall k \in N$. Now, since $M(y,\varepsilon_k) \subset M(\bar{y},\varepsilon_o)$, for all $k \in N$, and $M(\bar{y},\varepsilon_o)$ is compact, we may suppose without loss of generality that the sequence $\{w^k\}_N$ is convergent to, say, $\bar{w}$. Then, the upper-continuity of $M(y,\cdot)$ at $\varepsilon = 0$, under hypothesis i) and ii), asserts that $\bar{w} \in M(\bar{y},0)$, thus $\phi \neq M(\bar{y},0) \subset M(\bar{y},\varepsilon) \subset M(\bar{y},\varepsilon_o)$. The result then follows from the fact that $M(\bar{y},\varepsilon)$ is always closed under the hypothesis i) and ii).

Lemma 2. Let W be a finite dimensional normed space, Y and arbitrary metric space and let $\Omega: Y \rightarrow\!\!\triangleright W$, $f: W \times Y \to \bar{\mathbb{R}}$ verify, at a given point $\bar{y} \in Y$:

i) $\Omega(\bar{y})$ is a closed convex subset of W,
ii) $f(\cdot,\bar{y})$ is upper-semicontinuous over $\Omega(\bar{y})$,
iii) $f(\cdot,\bar{y})$ is quasiconcave over $\Omega(\bar{y})$.

Then, the partial point-to-set map $M(\bar{y},\cdot)$, defined over $\mathbb{R}_+$, is closed convex valued and upper-continuous over $\mathbb{R}_+$ and verifies: For any given $\alpha \geq 0$, $M(\bar{y},\alpha)$ is nonempty and compact if and only if there exists and $\varepsilon_o > 0$ such that $M(\bar{y},\alpha+\varepsilon)$ nonempty and compact for all $\varepsilon \in [-\alpha,\varepsilon_o]$.

Proof: The convexity of $M(\bar{y},\varepsilon)$, for all $\varepsilon \in \mathbb{R}_+$, is a straightforward consequence of the convexity of $\Omega(\bar{y})$ and the quasiconcavity of $f(\cdot,\bar{y})$

over $\Omega(\bar{y})$. The first part is then a consequence of part a) of Lemma 1. We then prove the second part. If $v(\bar{y}) = +\infty$, the result is trivially true. If $v(\bar{y}) < +\infty$ and if $M(\bar{y},\alpha)$ is nonempty, $M(\bar{y},\alpha+\varepsilon)$ is trivially nonempty for all $\varepsilon \in \mathbb{R}_+$. Consequently, if there is no such $\varepsilon_o > 0$, there must be sequences $\{\varepsilon_k \; \mathbb{R}_{++}\}_N$, $\{w^k \in \Omega(\bar{y})\}_N$ such that $\varepsilon_k \to 0_+, w^k \in M(\bar{y},\alpha+\varepsilon_k)$, $\forall k \in N$, $||w^k|| \to +\infty$, with $||w^k|| > 0$, $\forall k \in N$. Now, let $\bar{w}$ be any but fix, point in $M(\bar{y},\alpha)$ and let $\rho > 0$ be an arbitrary positive number. If we define $u^k =: (1-\lambda_k)\bar{w} + \lambda_k w^k$, $\forall k \in N$, where $\lambda_k =: \rho / ||w^k||$, there exists $\bar{k}$ such that $\lambda_k \in [0.1]$, $\forall k \geq \bar{k}$. Then, from the convexity of $M(\bar{y},\alpha+\varepsilon_k)$ and the fact that $\bar{w}$ also belongs to $M(\bar{y},\alpha+\varepsilon_k)$ we have that $u^k \in M(y,\alpha+\varepsilon_k)$, $\forall k \geq \bar{k}$. On the other hands, since $(w^k/ ||w^k||) \in S(0,1)$, $\forall k \in N$, and $S(0,1)$ is compact in W, we may suppose, by subsequencing if necessary, that $w^k/ ||w^k|| \to ||\bar{d}||$, with $||\bar{d}|| = 1$. Then, as $\lambda_k \to 0$, $\{u^k\}_{k \in N}$ is convergent to $(\bar{w}+\rho\bar{d})$. Then, from the upper-continuity of $M(\bar{y},\cdot)$ at $\varepsilon = \alpha$, we have that $(\bar{w}+\rho\bar{d}) \in M(\bar{y},\alpha)$. But since this result is true for any $\rho > 0$, $M(y,\alpha)$ cannot be bounded, thus a contradiction.

An immediate consequence of Lemma 2 is:

<u>Property 1:</u> Let X be a finite dimensional normed space, C a closed convex subset of X, and $\phi: C \to \mathbb{R}^m$, verify:

i) $\phi_i(\cdot)$ is quasiconvex over C,

ii) $\phi_i(\cdot)$ is lower-semicontinuous over C,

for all $i \in \{1,\dots, m\}$, and let $\Phi(\alpha) =: \{x \in C / \phi_i(x) \leq \alpha_i, i = 1,\dots,m\}$, for any $\alpha \in \mathbb{R}^m$.

Then, $\Phi(\bar{\alpha})$ is non-empty and compact for some $\bar{\alpha} \in \mathbb{R}^m$ if and only if there exists an $\alpha^o \in \mathbb{R}^m$, $\alpha^o > \bar{\alpha}$ (i.e. $\alpha_i^o > \bar{\alpha}_i$, $\forall i \in \{1,\dots,m\}$) such that $\Phi(\alpha)$ is nonempty and compact for all $\alpha \in [\bar{\alpha}\; \alpha^o]$, where $[\alpha\; \alpha^o] =: \{\alpha \in \mathbb{R}^m: \bar{\alpha}_i \leq \alpha_i \leq \alpha_i^o, \forall i \in \{1,\dots,m\}\}$.

<u>Proof:</u> Necessary Condition:

Let us first notice that, without loss of generality, we may consider that $\bar{\alpha} = 0$. In fact, $\Phi(\bar{\alpha}) = \bar{\Phi}(0) =: \{x \in C \;/\; \bar{\phi}_i(x) \leq 0,\; i = 1,\dots,m\}$, where the functions $\bar{\phi}_i(x) =: \phi_i(x) - \bar{\alpha}_i$, $\forall i \in \{1,\dots,m\}$, naturally verify the hypothesis i) and ii) for $\phi_i(\cdot)$, $i \in \{1,\dots,m\}$. We may then equivalently prove the existence of some strictly positive scalar $\varepsilon_o > 0$ such that $\bar{\Phi}(\varepsilon)$ is (nonempty) and compact for all $\varepsilon \in [0,\varepsilon_o]$. Moreover, since $\bar{\Phi}(\varepsilon) = \{x \in C / \theta(x) \leq \varepsilon\}$, where $\theta(\cdot): C \to \mathbb{R}$, defined by $\theta(x) =: \max_{i=1,\dots,m} \phi_i(x)$, verifies the same properties i) and ii) as each $\phi_i(\cdot)$, we may limit ourselves to prove the result for $\bar{\alpha} = 0$ and $m = 1$.

Let $\Phi(0)$ be nonempty and compact, where $\Phi(\varepsilon) = \{x \in C / \phi(x) \leq \varepsilon\}$, $\phi: C \to \mathbb{R}$. Then, from the hypothesis ii) and the closedness of C, there exists an $\bar{x} \in C$ such that $\phi(\bar{x}) = \inf\{\phi(x) \mid x \in C\}$, with $\phi(\bar{x}) \leq 0$ and $\phi(\bar{x}) > -\infty$ (Why?). Consequently, considering $W = X$, $f(x,y) = \phi(x)$, $\forall x \in W$, $\forall y \in Y$ and $\Omega(y) = C$, $\forall y \in Y$, we have that $v(\bar{y}) = \phi(\bar{x})$, for any $\bar{y} \in Y$, so that $\Phi(\varepsilon) = M(\bar{y}, \bar{\varepsilon}+\varepsilon)$, for $\bar{\varepsilon} =: -\phi(\bar{x})$, for any $\bar{y} \in Y$ and any $\varepsilon \in \mathbb{R}_+$. Then, if $\Phi(0)$ is nonempty and compact, from Lemma 2 there exists and $\varepsilon_o > 0$ such that $\Phi(\varepsilon)$ is nonempty and compact for all $\varepsilon \in [0, \varepsilon_o]$, and the necessary condition follows. Finally, since the sufficient condition is trivially true, the proof is over.

<u>Lemma 3</u> : Let W be an arbitrary normed space, Y an arbitrary metric space and $\Omega : Y \twoheadrightarrow W$, a point-to-set map such that:

i) $\Omega(y)$ is a closed convex subset of W, $\forall y \in Y$,

ii) $\Omega(\cdot)$ is continuous at $\bar{y}$ i.e. upper and lower-continuous at $\bar{y}$ or, more generally, upper-continuous at $\bar{y}$ and lower semicontinuous at $(\bar{y}, \{\bar{w}\})$, for some point $\bar{w} \in \Omega(\bar{y})$ or, simply, at $(\bar{y}, \Omega(\bar{y}))$.

Then, the recession cone map $0^+\Omega : Y \twoheadrightarrow W$ is upper-continuous at $\bar{y}$.

<u>Proof</u>: Let $\{y^k \in Y\}_N$ be any sequence in Y converging to $\bar{y}$, $\{d^k \in W\}_N$ a sequence converging to $\bar{d}$, where $d^k$ is an infinite direction for the closed convex set $\Omega(y^k)$, $\forall k \in N$. Let $\bar{w} \in \Omega(\bar{y})$ be the point in $\Omega(\bar{y})$ such that $\Omega(\cdot)$ is lower-semicontinuous at $(\bar{w}, \Omega(\bar{y}))$. Then, by definition, there exists a subsequence $\{w^k \in \Omega(y^k)\}_{N'}$ such that $w^k \to \bar{w}$. Now, let $\rho > 0$ be an arbitrary positive number. Then, by i) and the definition of $d^k$, $(w^k + \rho d^k) \in \Omega(y^k)$, $\forall k \in N'$, so that, by the upper-continuity of $\Omega(\cdot)$ at $\bar{y}$, $(\bar{w} + \rho\bar{d}) \in \Omega(\bar{y})$. But this is true for any $\rho \in \mathbb{R}_+$ thus $\bar{d} \in 0^+\Omega(\bar{y})$ and the result then follows.

<u>Lemma 4</u> : Let W be a finite dimensional normed space, Y an arbitrary metric space and let $\Omega : Y \twoheadrightarrow W$ and $f : W \times Y \to \bar{\mathbb{R}}$ verify, at a given point $\bar{y} \in Y$:

i) $\Omega(\bar{y})$ is a nonempty closed convex subset of W,

ii) $f(\cdot, \bar{y})$ defines a positively homogeneous concave function over W, lower-semicontinuous over $\Omega(\bar{y})$ and such that $f(\tilde{w}, \bar{y}) > -\infty$ for some $\tilde{w} \in \Omega(\bar{y})$.

Then, for any given $\varepsilon \geq 0$, $M(\bar{y}, \varepsilon)$ is nonempty and compact if and only if $f(u, \bar{y}) < 0$, $\forall u \in 0^+\Omega(\bar{y})$, $u \neq 0$.

<u>Proof</u>: Let $\bar{\varepsilon} \geq 0$ be given. Then, the condition is trivially necessary.

In fact, if there were some $\bar{u} \in 0^+\Omega(\bar{y}), \bar{u} \neq 0$, such that $f(\bar{u},\bar{y}) \geq 0$, then, for any $\bar{w} \in M(\bar{y},\bar{\varepsilon})$ we would have that $f(\bar{w} + \rho\bar{u},\bar{y}) \geq f(\bar{w},\bar{y}) + \rho f(\bar{u},\bar{y}) \geq v(\bar{y}) - \bar{\varepsilon}$ or, equivalently, $(\bar{w} + \rho\bar{u}) \geq M(\bar{y},\bar{\varepsilon})$, $\forall \rho \in \mathbb{R}_+$, thus contradicting the boundedness of $M(\bar{y},\bar{\varepsilon})$. Conversely, let us suppose that $f(u,\bar{y}) < 0$, $\forall u \in 0^+\Omega(\bar{y})$, $u \neq 0$, where this condition is verified by emptyness if $0^+\Omega(\bar{y}) = \{0\}$. Now, since $f(\cdot,\bar{y})$ is upper-semicontinuous over $\Omega(\bar{y})$, if $M(\bar{y},0)$ is empty or unbounded, there exists an unbounded sequence $\{w^k \in \Omega(\bar{y})\}_N$ such that $\|w^k\| \to +\infty$ and $\underline{\lim}_{k\to+\infty} f(w^k,\bar{y}) \geq v(\bar{y}) - \bar{\varepsilon}$. Moreover, by subsequencing if necessary, we may suppose, without loss of generality, that $(w^k/\|w^k\|) \xrightarrow[k\to+\infty]{} \bar{d}$, for some $\bar{d} \in S(0,1)$. Next, for any given positive scalar $\rho > 0$, let us consider the sequence of points $u^k \equiv (1-\lambda_k)\bar{w} + \lambda_k w^k$, for some fixed point $\bar{w} \in \Omega(\bar{y})$, where $\lambda_k =: \rho/\|w^k\|$, $\forall k \in N$. Then, from the convexity of $\Omega(\bar{y})$, we have that $u^k \in \Omega(\bar{y})$, for all k sufficiently large. Moreover, from the closedness of $\Omega(\bar{y})$, $\lim_{k\to+\infty} u^k = (\bar{w} + \rho\bar{d}) \in \Omega(\bar{y})$. We have then proved that $\bar{d} \in 0^+\Omega(\bar{y})\setminus\{0\}$. But this situation is not possible if $0^+\Omega(\bar{y}) = \{0\}$. On the contrary, if this is not the case, from our initial hypothesis we have $f(\bar{d},\bar{y}) < 0$. In this case, from the upper-semicontinuity of $f(\cdot,\bar{y})$, there exists a $\bar{k} \in N$ such that $f(w^k/\|w^k\|,\bar{y}) < \frac{\theta}{2} < 0, \forall k \geq \bar{k}$, where $\theta =: f(\bar{d},\bar{y}) < 0$. From this, we have that:

$$v(\bar{y}) - \bar{\varepsilon} \leq \underline{\lim}_{k\to\infty} f(w^k,\bar{y}) = \lim_{k\to+\infty} (\|w^k\| f(w^k/\|w^k\|,\bar{y})) = -\infty ,$$

but this is impossible if there exists at least one $\tilde{w} \in \Omega(\bar{y})$ such that $f(\tilde{w},y) > -\infty$.

<u>Lemma 5</u> : Let W be a finite dimensional normed space, Y an arbitrary metric space and let $\Omega : Y \longrightarrow W$ and $f: W \times Y \to \bar{\mathbb{R}}$ verify, at a given $\bar{y} \in Y$:

i) $\Omega(\bar{y})$ is a closed convex subset of W,
ii) $f(\cdot,\bar{y})$ is upper-semicontinuous over $\Omega(\bar{y})$,
iii) $f(\cdot,\bar{y})$ is a concave function over $\Omega(\bar{y})$.

Then, the partial point-to-set map $M(\bar{y},\cdot)$, defined over $\mathbb{R}_+$, is closed convex valued and upper-continuous over $\mathbb{R}_+$, and verifies:

For any given $\alpha \geq 0$, $M(\bar{y},\alpha)$ is nonempty and compact if and only if $M(\bar{y},\alpha+\varepsilon)$ is nonempty and compact for all $\varepsilon \in [-\alpha,+\infty[$

<u>Proof</u>: If $v(\bar{y}) = +\infty$, the result is trivially true. If this is not the case, taking into account Lemma 1 , we may limit ourselves to prove the result for $\alpha = 0$. Let then $M(\bar{y},0)$ be nonempty and compact. If $+\infty > v(\bar{y}) > -\infty$, $M(\bar{y},\varepsilon)$ may be seen as the cartesian projection over

W of the $\varepsilon$-optimal solution set of the augmented equivalent convex problem with linear objective function in $\mathbb{R} \times W$, Max $\{\mu | \mu \leq f(x,\bar{y}), x \in \Omega(\bar{y})\}$. Then, if $v(\bar{y}) \in \mathbb{R}$, the result follows directly from Lemma 4. Finally, if $v(\bar{y}) = -\infty$, the result is trivially true since $M(\bar{y},0) = \Omega(\bar{y})$ in this case.

Property 2: Let X be a finite dimensional normed space, C a closed convex subset of X and $\phi: C \to \mathbb{R}^m$, verify:

i) $\phi_i(\cdot)$ is convex over C,

ii) $\phi_i(\cdot)$ is lower-semicontinuous over C.

for all $i \in \{1,\ldots,m\}$, and let $\Phi(\alpha) =: \{x \in C / \phi_i(x) \leq \alpha_i, i=1,\ldots,m\}$, for any $\alpha \in \mathbb{R}^m$.

Then, $\Phi(\bar{\alpha})$ is non-empty and compact for some $\bar{\alpha} \in \mathbb{R}^m$ if and only if $\Phi(\alpha)$ is non-empty and compact for all $\alpha \in \mathbb{R}^m$ such that $\alpha \geq \bar{\alpha}$.

Proof: As in the proof of Property 1, we limit ourselves to the case where $m = 1$ and $\bar{\alpha} = 0$ and we prove that of $\Phi(0)$ is nonempty and compact then $\Phi(\varepsilon)$ is nonempty and compact for any $\varepsilon \geq 0$, where $\Phi(\varepsilon) = \{x \in C | \phi(x) \leq \varepsilon\}$. But as it was proved there, $\Phi(\varepsilon)$ is the $(\bar{\varepsilon}+\varepsilon)$-optimal solution set for the convex problem: Min$\{\phi(x)/x \in C\}$, where $\bar{\varepsilon} = -\phi(\bar{x})$, for any $\bar{x} \in X$ such that $\phi(\bar{x}) = \inf\{\phi(x)/x \in C\}$. The result is then a straightforward consequence of Lemma 5.

Theorem 1: Let X be a finite dimensional normed space, Y an arbitrary metric space and let $\Omega: Y \longrightarrow X$, $f: X \times Y \to \bar{\mathbb{R}}$ verify:

i) $\Omega(y)$ is closed convex subset of X, for all $y \in N(\bar{y})$

ii) $\Omega(\cdot)$ is upper-continuous or closed at $\bar{y}$,

iii) $\Omega(\cdot)$ is lower-continuous at $\bar{y}$ or, more generally, if $M(\bar{y},0) \neq \phi$, lower-semicontinuous at $(\bar{y}, \{\bar{x}\})$, for some $\bar{x} \in M(\bar{y},0)$ [resp. lower-continuous at $\bar{y}$ or, more generally, lower-semicontinuous at $(\bar{y}, \{\bar{x}\})$, for some point $\bar{x} \in \Omega(\bar{y})$],

iv) $f(\cdot,y)$ is quasiconcave [resp. concave] in x over $\Omega(y)$, for any fixed $y \in N(\bar{y})$,

v) $f(\cdot,y)$ is upper-semicontinuous in x, over $\Omega(y)$, $\forall y \in N(\bar{y})$,

vi) $f(\cdot,\cdot)$ is continuous over $\Omega(\bar{y}) \times \{\bar{y}\}$, or, more generally, upper-semicontinuous over $\Omega(\bar{y}) \times \{\bar{y}\}$ and lower-semicontinuous at $(\bar{x},\bar{y})$, where $\bar{x}$ is as in the hypothesis iii), where $N(\bar{y})$ is some neighbourhood of $\bar{y}$.

Then, $M(\overline{y},0)$ is nonempty and compact if and only if $M(\cdot,\varepsilon)$ is nonempty and uniformly compact near $\overline{y}$, for all $0\leq\varepsilon\leq\varepsilon_o$, for some $\varepsilon_o>0$ [resp. for all $0\leq\varepsilon<+\infty$ ].

Proof: Let $M(\overline{y},0)$ be a nonempty and compact. Then, as a consequence of Lemma 2 [resp. Lemma 5], $M(\overline{y},\varepsilon)$ is nonempty and compact for all $0\leq\varepsilon\leq\varepsilon_o$, for some $\varepsilon_o>0$ [resp. for all $0<\varepsilon<+\infty$]. We choose now any $\varepsilon\in[0,\varepsilon_o]$ [resp. any $\varepsilon\geq 0$].

If the conclusion were false, there would exist a sequence $\{y^k\in N(\overline{y})\}_{k\in N}$, converging to $\overline{y}$, such that $M(y^k,\varepsilon)$ is empty, $\underset{j>k}{\cup} M(y^j,\varepsilon)$ is unbounded, or both situations occur for infinitely many k. Let $\overline{x}\in\Omega(\overline{y})$ verify the hypothesis iii). There exists then a subsequence of $\{y^k\}_{k\in N}$, that, without loss of generality, we consider to be $\{y^k\}_{k\in N}$ itself, and a sequence $\{x^k\in\Omega(y^k)\}_{k\in N}$ such that $x^k \xrightarrow[k\to+\infty]{} \overline{x}$. Let $\{\delta_j\in\mathbb{R}_+\}_{j\in N}$ be a sequence such that $\delta_j \xrightarrow[j\to+\infty]{} \varepsilon$ with $\delta_j>\varepsilon$, $\forall j\in N$. Then, if $v(y^k)<+\infty$, there exists a sequence $\{w^{j,k}\in\Omega(y^k)\}_{j\in N}$ such that $v(y^k)\leq f(w^{j,k},y^k)+\delta_j$, $\forall j\in N$. Moreover, if $M(y^k,\varepsilon)$ is empty, from the hypothesis v), this sequence necessarily verifies $\|w^{j,k}\| \xrightarrow[j\to+\infty]{} +\infty$. Otherwise, any accumulation point w would be in $M(y^k,\varepsilon)$. On the other hand, if $v(y^k)=+\infty$, there exists a sequence $\{w^{j,k}\in\Omega(y^k)\}_{j\in N}$ such that $\lim_{j\to+\infty} f(w^{j,k},y^k)=+\infty$, with $f(x^k,y^k)<f(w^{j,k},y^k)$, for all $j\in N$. Moreover, if in this case $M(y^k,\varepsilon)$ is empty, this sequence necessarily verifies $\|w^{j,k}\| \xrightarrow[j\to+\infty]{} +\infty$, as a consequence of the hypothesis v).

In this way, if $M(y^k,\varepsilon)$ is empty, or if $\underset{\ell>k}{\cup} M(y^\ell,\varepsilon)$ is unbounded, or if both situations occur for infinitely many k, there exists a subsequence $\{y^k\}_{k\in N'}$ and a sequence $\{w^k\in\Omega(y^k)\}_{k\in N'}$, such that $f(x^k,y^k)\leq f(w^k,y^k)+\delta_k$ (*), for all $k\in N'$, where $\delta_k$ converges to $\varepsilon$ when $k\to+\infty$, and such that $\|w^k\| \xrightarrow[k\to+\infty]{k\in N'} +\infty$, with $\|w^k\|>0$, $\forall k\in N'$. Now, let $\rho>0$ be an arbitrary positive number, and consider the sequence $\{u^k\}_{k\in N'}$ defined as: $u^k=(1-\lambda_k)x^k+\lambda_k w^k$, $\forall k\in N'$, where $\lambda_k=\rho/\|w^k\|$, $\forall k\in N'$. There exists then some $\tilde{k}\in N'$ such that $\lambda_k\in[0,1]$, $\forall k\geq\tilde{k}$ and, consequently, from the convexity of $\Omega(y^k)$, $u^k\in\Omega(y^k)$, $\forall k\in N'$, $k\geq\tilde{k}$. Moreover, from the quasiconcavity of $f(\cdot,y^k)$ over $\Omega(y^k)$ and the inequality (*), we have that $f(x^k,y^k)\leq f(u^k,y^k)+\delta_k$, $\forall k\geq\tilde{k}$ (**). On the other hand, by subsequencing if necessary, we may suppose that $\{w^k/\|w^k\|\}_{k\in N'}$ converges to $d\in S(0,1)$, so that $u^k \xrightarrow[k\to+\infty]{} (\overline{x}+\rho\overline{d})$. Then, from the upper-continuity of $\Omega(\cdot)$ at $\overline{y}$, we have $(\overline{x}+\rho\overline{d})\in\Omega(\overline{y})$ so that, passing to the limit in the inequality (**), taking into account the continuity hypothesis vi) for $f(\cdot,\cdot)$, it results that $f(\overline{x},\overline{y})\leq f(\overline{x}+\rho\overline{d},\overline{y})+\varepsilon$. From the definition of $\overline{x}$, we have then that $v(\overline{y})\leq f(\overline{x}+\rho\overline{d},\overline{y})+\varepsilon$ [resp. $v(\overline{y})\leq f(\overline{x}+\rho\overline{d},\overline{y})+\overline{\varepsilon}$, where $\overline{\varepsilon}$ is defined by

$\bar{\bar{\varepsilon}} = \varepsilon + v(\bar{y}) - f(\bar{x},\bar{y}) > 0$, if $v(\bar{y}) < +\infty$, or $f(\bar{x}+\rho\bar{d},\bar{y}) = +\infty$, if $v(\bar{y}) = +\infty$, since $f(x) = +\infty$, $\forall x \in \Omega(\bar{y})$, in this case (Why?)] with $(\bar{x}+\rho d) \in \Omega(y)$, for all $\rho \geq 0$. But, since $\bar{d} \neq 0$, this contradicts the boundeness of $M(\bar{y},0)$ [resp. of $M(\bar{y},\bar{\varepsilon})$], and the result follows.

Remark: In Theorem 1,iii) and vi) may be replaced by: $\Omega(\cdot)$ is lower-semicontinuous at $(\bar{y},M(\bar{y}))$[resp.at$(\bar{y},\Omega(\bar{y}))$] and $f(\cdot,\cdot)$ is continuous over $M(\bar{y})x\{\bar{y}\}$ [resp.$\Omega(\bar{y})x\{\bar{y}\}$] or, alternatively, by: $f(\cdot,\cdot)$ is upper-semicontinuous over $\Omega(\bar{y})x\{\bar{y}\}$ and the general hypothesis H): $\forall\{y^k|k\in N\}\to\bar{y}, \exists\ \bar{x}\in M(\bar{y},0)$ [resp. $\bar{x}\in\Omega(\bar{y})$], $N'\subset N$, $|N'|=+\infty$, $\{x^k\in\Omega(y^k)|k\in N'\}$ such that $x^k \underset{k\to+\infty}{\to} \bar{x}$ and $\lim\limits_{k\to\infty} f(x^k,y^k) = f(\bar{x},\bar{y})$ [resp. $\lim\limits_{k\to\infty} f(x^k,y^k) > -\infty$]. On the other hand, if $f(\cdot,\bar{y})$ is quasiconcave and sup-bounded over $\Omega(\bar{y})$, the same result as for the concave case given in Theorem 1, is true.

Corollary 1.1: If X is a finite dimensional normed space, Y an arbitrary metric space and $\Omega: Y\to X$, a point-to-set map verifying, at given point $\bar{y}\in Y$:

i) $\Omega(\cdot)$ is upper-continuous at $\bar{y}$,

ii) $\Omega(\cdot)$ is lower-continuous at $\bar{y}$ or, more weakly, lower-semicontinuous at $(\bar{y},\Omega(\bar{y}))$,

iii) $\Omega(y)$ is a convex subset of X, for all y in some neighborhood N(y) of $\bar{y}$.

Then, $\Omega(\bar{y})$ is nonempty and compact if and only if $\Omega(\cdot)$ is nonempty and uniformly compact near $\bar{y}$.

Proof.: If we consider the map $\bar{\Omega}(\cdot)$, the topological closure of $\Omega(\cdot)$, and $f(x,y)=K$, $\forall(x,y)\in XxY$, some constant map, $\Omega(\cdot)$ and $f(\cdot,\cdot)$ satisfy all the hypothesis of Theorem 1. The result follows then from the fact that $M(\bar{y})=\bar{\Omega}(\bar{y})$ (since $\Omega(\bar{y})$ is closed) and $M(y) = \bar{\Omega}(y)$, $\forall y\in N(\bar{y})$, taking into account the above remark.

Corollary 1.2: Let X be a finite dimensional normed space, C a nonempty convex closed subset of X, Y an arbitrary metric space and let $f: CxY \to \mathbb{R}$ be such that, at some point $\bar{y}\in\bar{Y}$:

i) $f(\cdot,\bar{y})$ is inf-bounded in x over C,

ii) $f(\cdot,\bar{y})$ is convex in x over C, $\forall y\in N(\bar{y})$,

iii) $f(\cdot,y)$ is lower-semicontinuous in x, over C, $\forall y\in N(\bar{y})$,

iv) $f(\cdot,\cdot)$ is continuous over $\Omega(\bar{y})x\{\bar{y}\}$, where N(y) is some neighborhood of $\bar{y}$. Then, $f(\cdot,y)$ is inf-compact in x over C, for all y in some neighborhood of $\bar{y}$.

Proof: Since C is nonempty and closed and $f(\cdot,\bar{y})$ is inf-compact in x over C, the set $M(\bar{y},0) =: \{x\in C/f(x,\bar{y}) = v(\bar{y})\}$ ,where $v(y)=\inf\{f(x,y)/x\in C\}$, is nonempty and compact. Then, from Theorem 1, with $\Omega(y)=C$ for all $y\in N(\bar{y})$, there exist a neighborhood $N(\bar{y})$ of $\bar{y}$ such that $M(y,\varepsilon)$ is nonempty and compact for all $y\in N(\bar{y})$ and for all $\varepsilon\in\mathbb{R}_+$, thus completing the proof.

Theorem 2: Let X be a finite dimensional normed space, C a nonempty closed convex subset of X, Z any compact metric space and $f: CxZ \to \mathbb{R}$, a numerical function verifying:

i) $f(x,z) \xrightarrow[\|x\|\to+\infty]{x\in C} +\infty$ i.e. $f(\cdot,z)$ is inf-bounded in x over C, for any fixed $z\in Z$.

ii) f is continuous over $C\times Z$.

iii) $f(\cdot,z)$ is quasiconvex in x over C, for any fixed $z\in Z$.

Then, there exists at least one pair (x,z) minimizing f(x,z) over $C\times Z$, and the set of all the $\varepsilon$-minimizing pairs is compact in $X\times Z$, for any $\varepsilon\geq 0$.

Proof: Let $\{(x^k,z^k)\in C\times Z\}_{k\in N}$ be a minimizing sequence for $f(\cdot,\cdot)$ over $C\times Z$ i.e. a sequence such that $\lim_{k\to+\infty} f(x^k,z^k) = \Delta =: \inf\{f(x,z)/(x,z)\in C\times Z\}$. Since Z is compact, we may suppose, by subsequencing if necessary, that $\{z^k\}_{k\in N}$ is convergent to, say, $\bar{z}$. Now, let us consider the point-to-set map $M:Z\to X$ defined by: $M(z) =: \{x\in C/f(x,z)\leq v(z)\}$, where, $v(z)=: \sup\{f(x,z)/x\in C\}$. Since C is closed and $f(\cdot,z)$ inf-compact over C, $M(\bar{z})$ is clearly nonempty and compact. Then, from Theorem 1, with $Y=Z$ and $\Omega(\cdot) = C$ (the constant map such that $\Omega(z) = C, \forall z\in Z$), $M(\cdot,\varepsilon)$ is nonempty and uniformly compact near $\bar{z}, \forall\,\varepsilon\geq 0$. There exists then a subsequence $\{z^k\}_{k\in N'}$ and a sequence $\{w^k\in M(z^k)\}_{k\in N'}$ such that $w^k \xrightarrow[k\to+\infty]{N'} \bar{w}$, where $\bar{w}\in M(\bar{z})$ as a consequence of the upper-continuity of $M(\cdot)$ at $\bar{z}$. Then, since $x^k\in C$, for all $k\in N'$, and $f(\cdot,\cdot)$ is lower-continuous at $(\bar{w},\bar{z})$, we have that: $\Delta=\lim_{k\to\infty} f(x^k,z^k)\geq \varliminf_{k\to\infty} f(w^k,z^k)\geq f(\bar{w},\bar{z}), (\bar{w},\bar{z})\in C\times Z$. Then, the pair $(\bar{w},\bar{z})$ defines a minimizing point for $f(\cdot,\cdot)$ over $C\times Z$. Let us now show that $\{(x,z)\in C\times Z/f(x,z)-\varepsilon\leq\Delta\}$ is compact for any given $\varepsilon\geq 0$. If this were not the case, since Z is compact, there would exist a sequence $\{(x^k,z^k)\in C\times Z\}_{k\in N}$ such that $f(x^k,z^k)\leq\Delta+\varepsilon$, $\forall k\in N$, with $\|x^k\| \xrightarrow[k\to+\infty]{} +\infty$ and $z^k \xrightarrow[k\to+\infty]{} \tilde{z}\in Z$. But, since clearly $x^k\in M(x^k,\varepsilon)$, $\forall k\in N$, this is in contradiction with the fact that $M(\cdot,\varepsilon)$ is uniformly compact near z, and the result follows.

Remark: The following example shows that the result of Theorem 2 is not necessarily true if $f(\cdot,z)$ is not quasiconvex in x over $C, \forall z\in Z$. Let us consider $X=\mathbb{R}_+$, $Z=[0,1]$ and $f:\mathbb{R}_+\times Z\to\mathbb{R}$ defined by:

$$f(x,z) = \begin{cases} x; \text{ if } |x-1/z|\geq 1 \text{ or } z=0 \\ x[1-2e\exp(-1/(1-|x-1/z|))], \text{ otherwise.} \end{cases}$$

Then, we way easily see that $f(x,z)\to+\infty$ if $|x|\to+\infty$, for any fixed $z\in Z$. Nevertheless, for the sequence $(x^k,z^k)=:(k,1/k)\in X\times Z$, $\forall k\in\mathbb{N}$, we have that $\lim_{k\to+\infty} f(x^k,z^k) = -\infty$. This result is not in contradiction with Theorem 2 since it may be easily seen, in this case, that $f(\cdot,z)$ is not quasiconvex in $x$ over $X = \mathbb{R}_+$, for all $z \in Z$.

Theorem 3: Let $X,Y$ be arbitrary metric spaces and let $\Omega: Y \longrightarrow X$, $\Gamma: Y \longrightarrow X$ be the point-to-set maps defined as:

$$\Omega(y) = \{x \in D(y)/g(x,y) \leq 0\},$$

$$\Gamma(y) = \{x \in D(y)/g(x,y) < 0\},$$

where $D : Y \longrightarrow X$ is a given point-to-set map and $g: X \times Y \to \overline{\mathbb{R}}$, an extended real function. Then, if $\bar{y}$ is a given point in $Y$, we have the following results:

i) $D(\cdot)$ upper-continuous at $\bar{y}$; $g(\cdot,\cdot)$ lower-semicontinuous over $D(\bar{y}) \times \{\bar{y}\}$ $\Longrightarrow \Omega(\cdot)$ upper-continuous at $\bar{y}$.

ii) $D(\cdot)$ lower-continuous at $\bar{y}$ [resp. lower-semicontinuous at $(\bar{y},\Gamma(\bar{y}))$]; $g(\cdot,\cdot)$ upper-semicontinuous over $D(\bar{y}) \times \{\bar{y}\}$ [resp. upper-semicontinuous over $\Gamma(\bar{y}) \times \{\bar{y}\}$] $\Longrightarrow \Gamma(\cdot)$ lower-continuous at $\bar{y}$ [resp. lower-semicontinuous at $(\bar{y},\Gamma(\bar{y}))$].

iii) $D(\cdot)$ lower-continuous at $\bar{y}$; $g(\cdot,\cdot)$ upper-continuous over $D(\bar{y}) \times \{\bar{y}\}$; Hypothesis H) : $x \in \Omega(\bar{y}) \Longrightarrow x \in \overline{\Gamma}(\bar{y})$, where $\overline{\Gamma}(\bar{y})$ denotes the topological closure of $\Gamma(\bar{y})$ $\Longrightarrow \Omega(\cdot)$ lower-continuous at $\bar{y}$.

Proof: i) Let $\{y^k \in Y\}_{k \in N}$, $\{x^k \in \Omega(y^k)\}_{k \in N}$ be convergent to $\bar{y}$ and $\bar{x}$ respectively. Then, from the upper-continuity of $D(\cdot)$ at $\bar{y}$, $\bar{x} \in D(\bar{y})$ and, from the lower-semicontinuity of $g(\cdot,\cdot)$ over $D(\bar{y}) \times \{\bar{y}\}$, $g(\bar{x},\bar{y}) < \liminf_{k \to \infty} g(x^k,y^k) \leq 0$, thus $\bar{x} \in \Omega(\bar{y})$.

ii) Let $\{y^k \in Y\}_{k \in N}$ be convergent to $\bar{y}$. Then, from the lower-semicontinuity of $D(\cdot)$ at $(\bar{y},\Gamma(\bar{y}))$, there exists a subsequence $\{y^k \in Y\}_{k \in N'}$ and a sequence $\{x^k \in D(y^k)\}_{k \in N'}$ converging to some $\bar{x} \in \Gamma(\bar{y})$. But, since $\bar{x} \in \Gamma(\bar{y})$ $f(\bar{x},\bar{y}) < 0$ and $\bar{x} \in D(\bar{y})$. Then, by the upper-semicontinuity of $f(\cdot,\cdot)$ at $(\bar{x},\bar{y})$, there exists a $\bar{k} \in N'$ such that $f(x^k,y^k) < 0$, $\forall k \in N'$, $k \geq \bar{k}$, and the result follows.

iii) Let $\bar{x} \in \Omega(\bar{y})$ and let $\{y^k \in Y\}_{k \in N}$ be convergent to $\bar{y}$. From the hypothesis H), there exists a sequence $\{x^h \in \Gamma(\bar{y})\}_{h \in N}$ converging to $\bar{x}$. Now, from ii), $\Gamma(\cdot)$ is lower-continuous at $\bar{y}$. Then, for each $h \in N$, there exists a sequence $\{x^{h,k}/k \in N'\}$ converging to $x^h$, such that $x^{h,k} \in \Gamma(y^k) \subset \Omega(y^k)$, and $d_X(x^{h,k},x^h) < 1/h$, for all $k \geq \bar{k}(h)$, for a certain $\bar{k}(h) \in N$. Let us now define $i(h) =: \max \{\bar{k}(h')/h' = 1,\dots,h\}$.

Then, if $\{i(h)\}_{k \in N}$ is bounded i.e. $i(h) \leq i_o \in N$, for all $h \in N$, then $x^{h,k} \in \Omega(y^k)$, $\forall k \geq i_o$, $\forall h \in N$ and, at the same time:
$d_X(x^{h,k},\bar{x}) \leq d_X(x^{h,k},x^h) + d_X(x^h,\bar{x}) < 1/h + d_X(x^h,\bar{x})$, $\forall\; k \geq i_o$, $\forall h \in N$.

Hence, considering the diagonal sequence $\{w^k =: x^{k,k}\}_{k\in N}$, we have that $w^k \in \Omega(y^k), \forall k \geq i_o$ and $w^k \xrightarrow[k\to\infty]{k\in N} \bar{x}$, thus proving the result. On the other hand, if $\{i(h)\}_{h\in N}$ is not bounded, we have $i(h) \xrightarrow[h\to+\infty]{} +\infty$. Then, if we define the sequence $\{w^k\}_{k\in N}$ by, $w^k =: x^{h,k}$, $\forall k \in [i(h), i(h+1)[, \forall h \in N$, we have that $w^k \in \Omega(y^k), \forall k \geq i(1)$ and $0 \leq d_X(w^k,\bar{x}) = d_X(x^{h,k},\bar{x}) \leq 1/h + d_X(x^h,\bar{x})$ $\forall k \in [i(h), i(h+1)[, \forall h \in N$. Consequently, since $i(h) \xrightarrow[h\to+\infty]{} +\infty$ monotonically and $x^h \xrightarrow[h\to+\infty]{} \bar{x}$, we have that $\{w^k\}_{k\in N}$ is convergent and $\lim_{k\to+\infty} w^k = \bar{x}$, and the result follows.

<u>Remarks</u>: In theorem 3, there is no loss of generality in considering $g : X \times Y \to \bar{\mathbb{R}}$ instead of $g: X \times Y\ [-\bar{\mathbb{R}}, \bar{\mathbb{R}}]^m$, since this last case reduces to the former defining $g(x,y) = \max g_i(x,y); i=1,\dots,m$.

<u>Corollary 3.1</u> : Let X be a normed space, Y an arbitrary metric space and $\Omega: Y \to X$, the point to set map defined by $\Omega(y) = \{x \in D(y) / g_i(x,y) \leq 0, i=1,..,m\}$, where $D: Y \to X$ is a given point-to-set mat and $g_i : X \times Y \to \bar{\mathbb{R}}$, $i=1,\dots,m$, given extended real functions. Then, if $D(\cdot)$ is upper-continuous at $\bar{y}$ and $g_i(\cdot,\cdot)$ is lower-semicontinuous over $D(\bar{y}) \times \{\bar{y}\}$, $\forall i\{1,\dots,m\}$, the map $\Omega(\cdot)$ is upper-continuous at $\bar{y}$. On the other hand, if $D(\cdot)$ is lower-continuous [resp. continuous] at $\bar{y}$, $D(\bar{y})$ is a convex subset of X, $g_i(\cdot,\bar{y})$ is convex or strongly-quasiconvex in x over $D(\bar{y})$ and upper-semicontinuous [resp. continuous] over $D(\bar{y}) \times \bar{y}$, for each $i\{1,..,m\}$, then under the additional hypothesis:

H) $\exists \tilde{x} \in D(\bar{y})$: $g_i(\tilde{x},\bar{y}) < 0$, $\forall i \in \{1,\dots,m\}$, $\Omega(\cdot)$ is lower-continuous [resp. continuous] at $\bar{y}$.

<u>Proof</u>. The result is a straightforward consequence of Theorem 3, defining the function $g: X \times Y \to \bar{\mathbb{R}}$ by $g(x,y) =: \max\{g_i(x,y) / i=1,\dots,m\}$ and taking into account the fact that, under the hypothesis H), $\Omega(\bar{y}) \subset \bar{\Gamma}(\bar{y})$ [ 2 ].

<u>Corollary 3.2</u>: Let X be a normed space, Y an arbitrary metric space, $D: Y \to X$ a given point-to-set map and $B: Y \to X$, the point-to-set map defined by: $B(y) = B(\bar{x}(y), r(y))$, the closed ball centered at $\bar{x}(y)$ with radius $r(y)$, where $\bar{x}: Y \to X$, $r: Y \to \mathbb{R}_{++}$ are given functions. Then if $\Omega: Y \to X$ is defined by: $\Omega(y) =: D(y) \cap B(y)$, we have the following results:

I) If $D(\cdot)$ is upper-continuous at $\bar{y}$, $\bar{x}(\cdot)$ is continuous at $\bar{y}$ and $r(\cdot)$ is upper-semicontinuous at $\bar{y}$, the map $\Omega(\cdot)$ is upper-continuous at $\bar{y}$.

II) If $D(\cdot)$ is lower-continuous [resp. continuous] at $\bar{y}$, $D(\bar{y})$ is a convex subset of X, $\bar{x}(\cdot)$ is continuous at $\bar{y}$, $r(\cdot)$ is lower-semicontinuous [resp. continuous] at $\bar{y}$ and $D(\bar{y}) \cap B(\bar{y}) \neq \phi$ , the map $\Omega(\cdot)$ is lower-continuous [resp. continuous] at $\bar{y}$.

Proof.: The result is a straightforward consequence of Corollary 3.1.

Application : Let X be a normed space, Y an arbitrary metric space and $\Omega$: Y $\longrightarrow$ X the map defined by $\Omega(y) = D(y) \cap \{x \in X / g_i(x,y) \leq 0, \forall i = 1,\ldots,m\}$, where D : Y $\longrightarrow$ X is a given map, and consider the family of parametrized problems $P(y)$:

$$\begin{array}{l|} \text{Max } f(x,y) \\ x \in \Omega(y) \\ \hline \end{array}$$

where $f: X \times Y \to \mathbb{R}$. Now, if for each y we consider the Lagrange function associated with problem P(y) defined by $L(x,u,y) =: f(x,y) - \sum_{i=1}^{m} u_i g_i(x,y)$, for all $x \in X$, for all $u \in \mathbb{R}^m$ and for all $y \in Y$, we may alternatively consider the parametrized family of equivalent problems, defined by: $P(y)$: Sup $\{F(x,y)/x \in \Omega(y)\}$, where $F(x,y) =: \text{Inf}\{L(x,u,y)/u \in \mathbb{R}^m_+\}$, and the corresponding dual family, defined by: $\mathcal{D}(y)$: $\text{Inf}\{H(u,y)/u \geq 0\}$, where $H(u,y) =: \text{Sup}\{L(x,u,y) / x \in \Omega(y)\}$ with $H(u,y) = +\infty$, $\forall u \in \mathbb{R}^m_+$, if $\Omega(y) = \phi$. Then if $v(\cdot), M(\cdot,\cdot)$ denote the optimal value and the quasi-optimal solutions map for $P(\cdot)$ respectively, and $d(\cdot)$, $U(\cdot,\cdot)$, the corresponding maps for $\mathcal{D}(\cdot)$, we have the following result:

Lemma 6: Let X,Y be arbitrary metric spaces, D:Y $\longrightarrow$ X a given point-to-set map and $g_i$: $X \times Y \to \mathbb{R}$ (i=1,...,m), given numerical functions such that, at a given point $\bar{y} \in Y$:

i) $D(\cdot)$ is lower-continuous at $\bar{y}$,

ii) $g_i(\cdot,\cdot)$ is upper-semicontinuous at $(\bar{x},\bar{y})$, $\forall i$ $\{1,\ldots,m\}$,

iii) $\exists \bar{x} \in D(\bar{y})$: $g_i(\bar{x},\bar{y}) < 0$, $\forall i \in \{1,\ldots,m\}$.

Then, for any ball $B(\bar{x},R)$ (open or closed) centered at $\bar{x}$, there exists a neighborhood $V(\bar{y})$ of $\bar{y}$ such that: $\forall y \in V(\bar{y})$, $\exists x \in D(y) \cap B(\bar{x},R)$: $g_i(x,y) < 0$, $\forall i \in \{1,\ldots,m\}$.

Proof: For a given $B(\bar{x},R)$, we define $\Gamma(y) =: D(y) \cap \overset{\circ}{B}(\bar{x},R)$, for all $y \in Y$, where $\overset{\circ}{B}(\bar{x},R)$ denotes the interior of $B(\bar{x},R)$. Then, under the hypothesis i), the map $\Gamma$ is clearly lower-continuous at $\bar{y}$. Now, let $(\bar{x},\bar{y})$ verify hypothesis iii) and suppose that the result is not true. There exists then a sequence $\{y^k\}_{k \in N}$ converging to $\bar{y}$ such that: $\forall k \in N$, $\forall x \in \Gamma(y^k)$, $\exists i = i(x) \in \{1,\ldots,m\}$ : $g_i(x,y^k) \geq 0$.

Now, since $\bar{x} \in \Gamma(\bar{y})$ and $\Gamma(\cdot)$ is lower-continuous at $\bar{y}$, there exists $\bar{k} \in N$ and a sequence $\{x^k \in \Gamma(y^k)/k \geq \bar{k}\}$ converging to $\bar{x}$. Consequently, since m is finite, there exists a subsequence $\{(x^k,y^k)\}_{k\in N'}$, and a fix $i_o \in \{1,\ldots,m\}$, such that $g_{i_o}(x^k,y^k) \geq 0$, $\forall k \in N'$. Then, passing to the limit when $k \xrightarrow{k\in N'} +\infty$ we have that $g_{i_o}(\bar{x},\bar{y}) \geq 0$, thus contradincting the hypothesis iii), and the proof is over.

Theorem 4: If X is a finite dimensional normed space and if the family of parametrized problems $P(y)$ verifies:

i) D is continuous (i.e. upper and lower-continuous) at $\bar{y}$,

ii) $D(y)$ is a closed convex subset of X, $\forall y \in N(\bar{y})$,

iii) $g_i(\cdot,\cdot)$ is continuous over $D(\bar{y})\times\{\bar{y}\}$ and $g_i(\cdot,y)$ is lower-semicontinuous in x over $D(y), \forall y \in N(\bar{y})$, $\forall i\in\{1,\ldots,m\}$,

iv) $g_i(\cdot,y)$ is convex in x over $D(y)$, $\forall y \in N(\bar{y})$, $\forall i \in \{1,\ldots,m\}$,

v) $\exists \bar{x} \in D(\bar{y}):\ g_i(\bar{x},\bar{y})<0$; $\forall i \in \{1,\ldots,m\}$,

vi) $f(\cdot,\cdot)$ is continuous over $\Omega(\bar{y})\times\{\bar{y}\}$ and $f(\cdot,y)$ is upper-semicontinuous in x over $\Omega(y)$, $\forall y \in N(\bar{y})$,

vii) $f(\cdot,y)$ is concave in x over $\Omega(y)$, $\forall y \in N(\bar{y})$, where $N(\bar{y})$ is some neighborhood of $\bar{y}$, then, if $v(\bar{y})$ is finite, we have the following result: if $M(\bar{y},0)$ is nonempty and compact, $M(\cdot,\cdot)$ and $U(\cdot,\cdot)$ are nonempty and uniformly compact near $(\bar{y},0)$. Moreover, $v(\cdot)$ and $d(\cdot)$ are continuous at $\bar{y}$ and $v(y)=d(y)$ for all y in some neighborhood of $\bar{y}$ i.e. there is no duality gap for $P(y)$ and $D(y)$, for all y near $\bar{y}$. Finally, $M(\cdot,\cdot)$ and $U(\cdot,\cdot)$ are upper-continuous over $\{\bar{y}\}\times\mathbb{R}_+$ and lower-continuous (thus, continuous) over $\{\bar{y}\}\times\mathbb{R}_{++}$.

Proof: For all $y \in N(\bar{y})$, it is clear that $P(y)$ is a convex problem. On the other hand, if $B(\bar{x},R)$ is a given ball, from Lemma 6 there is a neighborhood $V(\bar{y})$ such that, for any $y \in V(\bar{y})$, there exists an $x \in D(y) \cap B(\bar{x},R)$ verifying $g_i(x,y) < 0, \forall i\in\{1,\ldots,m\}$. Consequently, if $v(\bar{y}) < +\infty$, from the Strong Duality Theorem and the Stability Theorem for convex programs [12,5], $U(y,0)$ is nonempty and compact and $v(y) = d(y) = H(u,y)$, $\forall u \in U(y,0)$, for all $y \in V(\bar{y})$. On the other hand, from Theorem 1 and Corrollary 3.1, $M(\cdot,\cdot)$ is nonempty and uniformly compact near $(\bar{y},0)$ and $v(\cdot)$ is continuous at $\bar{y}$, and finite in a neighborhood of $\bar{y}$. There exists then a closed ball $K, K \supset B(\bar{x},R)$, such that $M(y,0) \subset \text{int}\ (K)$ for all y near $\bar{y}$. In this way, the retriction of the domain $D(y)$ to $D(y)\cap K$ gives rise to a new family of parametrized convex problems with the same optimal value function near $\bar{y}$ and whose constraint domains have the same continuity properties at $\bar{y}$ as the original ones. Moreover, these new problems still verify the Slater Constraint Qualification for all y near $\bar{y}$. Consequently, for all $y \in V(y)$, the corresponding dual problems $\mathcal{D}_K(y)$ have a nonempty compact set of solutions, say $U_K(y,0)$, and are such that $v(y) = H_K(u,y)$, $\forall u \in U_K(y,0)$, where $H_K(\cdot,y)$ denotes the objective function of $\mathcal{D}_K(y)$. Now, we are going to show that $\mathcal{D}(y)$ and $\mathcal{D}_K(y)$ have identical sets of optimal solutions near $\bar{y}$. Let $\bar{u} \in U(y,0)$. Then, if $\bar{x} \in M(y,0)$, we clearly have that $H(\bar{u},y) = v(y) = L(\bar{x},\bar{u},y) \leq H_K(\bar{u},y) \leq H(\bar{u},y)$, thus $H_K(\bar{u},y) = v(y)$. Then, from the Weak Duality Theorem for the problems $P_K(y)$ and $\mathcal{D}_K(y)$ and the fact that $\bar{u} \in \mathbb{R}^m_+$, $\bar{u} \in U_K(y,0)$. Conversely, if $\bar{u} \in U_K(y,0)$, then $H_K(\bar{u},y) = v(y)$ or, equivalently, if $\bar{x} \in M(y,0)$, $L(\bar{x},\bar{u},y) = v(y)$. Now, if $\bar{u} \notin U(y,0)$, $H(\bar{u},y) > v(y)$ and there exists then $\tilde{x} \in D(y) \setminus K$ such that $L((\bar{x},\bar{u},y) < L(\tilde{x},\bar{u},y)$. Then, since $\bar{x} \in D(y) \cap \operatorname{int}(K)$, there exists $\tilde{\lambda} \in ]0,1[$ such that $x(\tilde{\lambda}) =: [(1-\tilde{\lambda})\bar{x} + \tilde{\lambda}\tilde{x}] \in D(y) \cap K$. Consequently, since $L(\cdot,\bar{u},y)$ is concave in x, we have that $L(x(\tilde{\lambda}),\bar{u},y) > L(\bar{x},\bar{u},y)$, with $x(\tilde{\lambda}) \in K \cap D(y)$, thus contradicting the definition of $H_K(\bar{u},\bar{y})$. We have then proved the equality $U(y,0) = U_K(y,0)$, for all y near $\bar{y}$. Next, from Corollary 3.2 and the continuity of $D(\cdot)$ at $\bar{y}$, $K \cap D(\cdot)$ is clearly continuous at $\bar{y}$. Then, since $L(x,u,y)$ is continuous over $\Omega(\bar{y}) \times \mathbb{R}^m \times \{\bar{y}\}$ (under iii) and vi)) and $K \cap D(\cdot)$, locally uniformly compact at $\bar{y}$, $H_K(u,y)$ is continuous over $\mathbb{R}^m \times \{\bar{y}\}$, as a consequence of Berge Theorem. Moreover, since $L(x,\cdot,y)$ is continuous in u over $\mathbb{R}^m$, for any $(x,y) \in X \times Y$, $H_K(\cdot,y)$ is lower-semicontinuous in u over $\mathbb{R}^m$, $\forall y \in Y$. Then, since $H_K(\cdot,y)$ is convex in u over $\mathbb{R}^m$, $\forall y \in Y$, we have that $U_K(\cdot,\cdot)$ is nonempty and uniformly compact near $(\bar{y},0)$, as a consequence of Theorem 1, and the same is true for $U(\cdot,\cdot)$, since $\phi \neq U(y,0) \subset U(y,\varepsilon) \subset U_K(y,\varepsilon)$, for all y near $\bar{y}$. Finally, $d(\cdot)$ is continuous at y since $v(y) = d(y)$ for all y near $\bar{y}$, as a consequence of the fact that $\mathcal{D}(y)$ and $\mathcal{D}_K(y)$ have the same optimal solutions near $\bar{y}$. In particular then, $M(\cdot,\cdot)$ and $U(\cdot,\cdot)$ are upper-continuous at $(\bar{y},\bar{\varepsilon})$, $\forall \bar{\varepsilon} \in \mathbb{R}_+$, as a consequence of Theorem 3, and lower-continuous (thus, continuous) at $(\bar{y},\bar{\varepsilon})$, $\forall \bar{\varepsilon} > 0$, as a consequence of Corollary 3.1, taking into account the continuity of $v(\cdot)$ and $d(\cdot)$ at $\bar{y}$ and the fact that the sets $\{x \in \Omega(\bar{y}) / v(\bar{y}) - f(x,\bar{y}) < \bar{\varepsilon}\}$ and $\{u \in \mathbb{R}^m_+ / H(u,\bar{y}) - d(\bar{y}) < \bar{\varepsilon}\}$ are nonempty for all $\bar{\varepsilon} > 0$.

Remark: Theorem 4 (as, Theorem 10 [ 2 ] ) extends the analogous result given by Hogan [ 7 , Lemma 2] for the case where $D(\cdot)$ is the closed convex constant point-to-set map (i.e. the case where $D(y) = D, \forall y \in Y$, where D is a fix closed convex subset of X) and under more stringent continuity hypothesis for the functions $f(\cdot,\cdot)$ and $g_i(\cdot,\cdot), \forall i \in \{1,..,m\}$. At the time, an extension of Theorem 10[ 2 ] to some general parametrized saddle-point problems has been given by Correa and Seeger in [ 3 ]

REFERENCES.

[ 1 ] BERGE, C., "Topological Spaces", Macmillan, Nre York, 1963.

[ 2 ] CONTESSE, L., "On the Continuity of Optimal Value Functions and of Optimal Solution Sets", Publication A.N.O. Nº 62, Université de Lille I, 1981.

[ 3 ] CORREA, R. and SEEGER,A., "Some Topological Properties of the Solution Sets of Parametrized Minimax Problems", to appear in Applied Mathematics and Optimization.

[ 4 ] EVANS J.P. and GOULD, F.J., "Stability in Nonlinear Programming", Operations Research, 18(1970), 107-118.

[ 5 ] GEOFFRION, A.M., "Duality in Nonlinear Programming: a Simplified Applications Oriented Development", SIAM Review, 13(1971), 1-37.

[ 6 ] HOGAN, W.W., "Point-to-Set Maps in Mathematical Programming", SIAM Review, Vol. 15(3) (1973).

[ 7 ] HOGAN, W.W., "Directional Derivatives for Extremal Value Functions with Applications to the Completly Convex Case", Operations Research 21(1973), 188-209.

[ 8 ] HUARD, P., "Point-to-Set Maps and Mathematical Programming", Mathematical Programming Study 10 (1979).

[ 9 ] PENOT J.P., "Continuity Properties of Performance Functions" in Optimization: Theory and Algorithms, edited by J.B.Hiriart Urruty, W. Oettli and J. Stoer, Marcel Dekker (1983),77-90.

[ 10 ] ROCKAFELLAR, T.,"Duality and Stability in Extremum Problems Involving Convex Functions", Pacific Journal of Mathematics 21(1967), 167-187.

# SOME PROPERTIES OF SEMISMOOTH AND REGULAR FUNCTIONS IN NONSMOOTH ANALYSIS.

Rafael Correa
Facultad de Ciencias Físicas y Matemáticas
Departamento de Matemáticas y Ciencias de la Computación
Universidad de Chile
Casilla 170 - Correo 3, Santiago, CHILE

Alejandro Jofré
Departamento de Matemáticas
Facultad de Ciencias
Universidad de La Serena
Casilla 599
La Serena - CHILE

ABSTRACT:

Given a real valued function f, defined on a locally convex topological space X, locally Lipschitzian, and Gateaux-differentiable on a dense subset D in X, we have studied under what hypotheses Charke's generalized gradient can be written as

$$\partial f(x) = \overline{co}\ \{w^*\text{-}\lim_{y \to x} \nabla f(y) /\ y \in D\},$$

It is shown that this formula is valid in particular when f is regular or semismooth. By using this characterization, some properties known to hold true in finite dimension are generalized and other new properties are established. In particular, a characterization of semismooth functions is given in terms of the continuity of the directional derivative. Finally, characterizations for the directional derivative and generalized gradient of marginal functions are obtained. In particular, Mifflin's result stating that lower-$C^1$ functions are semismooth is generalized.

Partially supported by Univ. de La Serena, under grant Nº130.2.07 and Fondo Nacional de Ciencias, under grant Nº 01273.

## 1. Preliminary definitions.

The class of nonsmooth functions defined on a locally convex topological vector space ($\ell$.c.t.v.s.) X and taking on values on the real line $\mathbb{R}$ that nowadays draws the gratest attention consists of the so-called locally Lipschitzian functions, which we now define:

Definition 1.1.: A function $f : X \to \mathbb{R}$ is said to be locally Lipschitzian if for all $x \in X$ there exist a neighbourhood V of x, a continuous semi-norm p over X, and a constant $\alpha \geq 0$, such that

$$|f(y) - f(z)| \leq \alpha p(y-z) \quad \text{for all } y,z \in V.$$

When f is a continuous convex (concave) function on X (a fortiori locally Lipschitzian) the directional derivative

$$f'(x;d) = \lim_{t \to 0^+} \frac{f(x+td) - f(x)}{t} \tag{1.1}$$

plays a fundamental role in the so-called convex analysis, mainly developed by J.J. Moreau [11] and R.T. Rockafellar [12] in the 60's, and which may be considered at the present time as the first chapter of the nonsmooth analysis.

The fundamental property of this directional derivative, besides its existence, is that as a function of the direction d is finite, positively homogeneous, convex (i.e. sublinear) when f is convex, and concave (i.e. suplinear) when f is concave.

From these properties of $f'(x;\cdot)$, we can define the subdifferential of a continuous convex function f at a point $x \in X$ as the nonempty w* compact convex set

$$\partial f(x) = \{x^* \in X^* \;/\; f'(x;d) \geq \langle x^*, d\rangle \text{ for all } d \in X\}$$

where X* is the topological dual of X and < , > represents the canonical bilinear form of the duality on X* x X. Analogously when f is a continuous concave function, the supdifferential of f at x is

$$\partial f(x) = \{x^* \in X^* \;/\; f'(x;d) \leq \langle x^*, d\rangle \text{ for all } d \in X\}$$

It can be proved, in both cases, that f will be Gateaux-differentiable at x if and only if the set $\partial f(x)$ reduces to just one point, which corresponds to the Gateaux derivative of f at x.

Outside the class of concave and convex functions, the directional derivative does not in general verify the aföresaid sub or suplinearity property and the results of convex analysis are not valid. The situation there is worse because $f'(x;d)$ might not even exists for a locally lipschitzian function.

In 1973 F.H. Clarke introduced a new directional derivative which allows generalizing the sup and subdifferential notions for a wider class of functions, namely the locally Lipschitzian functions.

Definition 1.2. : Let $f : X \to \mathbb{R}$ be locally Lipschitzian. The upper and lower generalized directional derivatives of f at $x \in X$ in the direction d, are defined by

$$f^\circ(x;d) = \limsup_{\substack{y \to x \\ t \to 0^+}} \frac{f(y+td)-f(y)}{t} \tag{1.2}$$

$$f_o(x;d) = \liminf_{\substack{y \to x \\ t \to 0^+}} \frac{f(y+td)-f(y)}{t}$$

It can be shown that $f^\circ(x;\cdot)$ is sublinear, that $f_o(x;\cdot)$ is suplinear, and that $f'(x;d)$ coincides with $f^\circ(x;d)$ or $f_o(x;d)$ when f is convex or concave respectively. These generalized directional derivatives trivially verify $f^\circ(x;d) = - f_o(x;-d)$.

By means of this generalized directional derivatives Clarke extends the notions of sup and subdifferential to the class of locally Lipschitzian functions.

Definition 1.3. : For a locally lipschitzian function $f : X \to \mathbb{R}$, the generalized gradient of f at $x \in X$ is defined as

$$\begin{aligned} \partial f(x) &= \{x^* \in X^* \,/\, f^\circ(x;d) \geq \langle x^*,d\rangle \text{ for all } d \in X\} \\ &= \{x^* \in X^* \,/\, f_o(x;d) \leq \langle x^*,d\rangle \text{ for all } d \in X\} \end{aligned} \tag{1.3}$$

It can be shown that $\partial f(x)$ is a nonempty w*-compact convex set in X*.

From this notion Clarke develops the nonsmooth analysis for locally Lipschitzian functions. A detailed explanation of this subject can be found in the excellent book [5] .

It is important to observe that unlike the sub and supdifferential notions, generalized gradient do not privilege convexity or concavity. Characterization (2.1) given in the next section clearly shows this idea.

If the generalized gradient of a locally Lipschitzian function at a point $x \in X$ is reduced to just one element, then the function will be Gateaux-differentiable at x. Unfortunately the reciprocal is not true, except when f is continuously Gateaux-differentiable at x. In general, we know that the Gateaux derivative of f at x belongs to the generalized gradient of f at x.

It is important to point out that this generalized gradient notion is not interesting for every locally Lipschitzian function. We mention in this respect that Clarke has built a Lipschitzian function on $\mathbb{R}$ whose generalized gradient at all points coincides with $[-1,1]$. In the next section, we will obtain some important properties for three subclasses of locally Lipschitzian functions: the subregular, supregular, and semismooth functions.

## 2. Regular and semismooth functions.

The starting point of the nonsmooth analysis (non-convex) was the definition given by Clarke [4] of the generalized gradient for a locally Lipschitzian real-valued function defined on $\mathbb{R}^n$. By using a result by Rademacher which stating that a real-valued function f which is locally Lipschitzian on $\mathbb{R}^n$ is Gateaux-differentiable almost everywhere (a.e) in $\mathbb{R}^n$ Clarke defined the generalized gradient of f at a point $x \in \mathbb{R}^n$ by

$$\partial f(x) = \mathrm{co}\{\lim_{y \to x} \nabla f(y) / y \in G\}, \qquad (2.1)$$

where co A denotes the convex hull of the set A, $\nabla f(y)$ denotes the Gateaux-derivative of f at y, and G is the set (whose complement is Lebesgue-null) where f is Gateaux-differentiable. Clarke proved that (2.1) and (1.3) coincide.

The generalization of (2.1), when f is defined on an infinite-dimensional space, previously requires a generalization of Rademacher's theorem. In this sense Christensen [3] proved that if X is a separable Banach space and f is a locally Lipschitzian function defined on X, then f is Gateaux-differentiable excepting on a Haar-null set (in $\mathbb{R}^n$, the Haar-null sets and Lebesgue-null sets coincide). Other results in this way have been obtained by Aronsajn [1] and Mignot [10]. By taking (1.3) as the definition of the generalized gradient, and using the result of Christensen, Thibault [16] obtained the characterization

$$\partial f(x) = \overline{co}\{w^*\text{-}\lim_{y\to x} \nabla f(y) / y \in G\}, \tag{2.2}$$

that is, the w*-closed convex hull of the w*-limits of Gateaux-derivatives $\nabla f(y)$ when y converges to x, with y in the set where f is Gateaux-differentiable.

In what follows we will investigate what occurs to characterization (2.2) when f is defined on a $\ell$.c.t.v.s. X and when we replace G by a dense subset D in X where f is Gateaux-differentiable.

When G is replaced by a dense subset D of X, characterization (2.2) is no longer valid in general, even if X is finite dimensional. As we said in section 1, there exists a Lipschitzian function $f:\mathbb{R}\to\mathbb{R}$ such that $\partial f(x) = [-1,1]$ for all $x \in \mathbb{R}$. In general, characterization (2.2) is valid if the complement of D is Haar-null.

The next theorem due to Shu-Chung [14], give a global answer to these two questions:

Theorem 2.1 : Let X be a $\ell$.c.t.v.s. and f a locally Lipschitzian function on X. If f is Gateaux-differentiable on a dense subset D in X, a necessary and sufficient condition for the equality

$$\partial f(x) = \overline{co}\ \{w^*\text{-}\lim_{y\to x} \nabla f(y) / \ y \in D\}, \tag{2.2}$$

is that for all $x,d \in X$ such that the bilateral directional derivative

$$Df(x)(d) = \lim_{t\to 0} \frac{f(x+td) - f(x)}{t}$$

exists, one has

$$Df(x)(d) \le \limsup_{\substack{y\to x\\ y\in D}} \langle \nabla f(y), d\rangle \qquad (2.3)$$

Remark 2.2 According to the hypothesis of Gateaux-differentiability of a locally Lipschitzian function f on a dense subset in a ℓ.c.t.v.s. X, Lebourg [8] proved that if X is a separable Baire space (for example a separable Banach space) and if f admits generically a directional derivative $f'(x;d)$ for every $d \in X$ (this will occur for supregular, subregular and semismooth functions), then f is generically Gateaux - differentiable (the term "generically" means here that the property holds on a dense $G_\delta$ subset in X). Another recent result in this sense has been given by Zivkov [17]; it says that if X is an H-space (in particular a reflexive Banach space) and if for all $x \in X$ the directional derivative $f'(x;d)$ exists for all d in a dense subset of X, then f is generically Gateaux-differentiable on X.

We now introduce three important classes of locally Lipschitzian functions defined on a ℓ.c.t.v.s. X.

Definition 2.3. A locally Lipschitzian function $f : X \to \mathbb{R}$ is subregular (respectively supregular) at $x \in X$ provided that for all $d \in X$ the directional derivative $f'(x;d)$ exists and $f'(x;d) = f^\circ(x;d)$ (respectively $f'(x;d) = f_o(x;d)$). A function f is subregular (respectively supregular) if it is so at every $x \in X$.

Remark 2.4. f is subregular at x if and only if $-f$ is supregular at x.

Remark 2.5. These classes of nonsmooth functions were introduced by Clarke [4] . The subregular functions are called regular by Clarke and subdifferentially regular by Rockafellar. Both classes are a convex cone and include the subspace of continuously differentiable functions. The subregular (respectively supregular) functions include the cone of continuous convex (respectively concave) functions.

Definition 2.6. A locally Lipschitzian function $f : X \to \mathbb{R}$ is semismooth at $x \in X$ if for all nets $t_\alpha$ in $\mathbb{R}_+$ converging to 0, $d_\alpha$ in X converging to d, and $x^*_\alpha$ in $X^*$ such that $x^*_\alpha \in \partial f(x+t_\alpha d_\alpha)$, one has $\lim_\alpha \langle x^*_\alpha, d\rangle = f'(x;d)$. The function f is semismooth if it is so at every $x \in X$.

Remark 2.7. : The semismooth functions were introduced by Mifflin [9] when $X = \mathbb{R}^n$. This class makes up a vector space including the subspace of continuously differentiable functions.

Proposition 2.8. If f is a continuous convex or concave function, then it is semismooth.

Proof. We suppose f continuous and convex, a fortiori locally Lipschitzian. Let $x \in X$, $t_\alpha$ converging to 0, $d_\alpha$ converging to d and $x^*_\alpha \in \partial f(x+t_\alpha d_\alpha)$ be given. By convexity

$$<x^*_\alpha, d_\alpha> \geq \frac{f(x+t_\alpha d_\alpha) - f(x)}{t_\alpha} \quad \text{for all } \alpha,$$

and since f is locally Lipschitzian and $f'(x;d)$ exists, it follows that

$$\liminf_\alpha <x^*_\alpha, d_\alpha> \geq f'(x;d).$$

On the other hand, since f is subregular we have $f'(x+t_\alpha d_\alpha; d_\alpha) \geq$ $<x^*_\alpha, d_\alpha>$, and taking upper limit we obtain from the u.s.c. of $f'(\cdot;\cdot)$ that

$$f'(x;d) \geq \limsup_\alpha <x^*_\alpha, d_\alpha>,$$

finally since $\partial f$ is uniformly equicontinuous in a neighbourhood of x, we conclude that $\lim_\alpha <x^*_\alpha, d> = f'(x;d)$

□

Corallary 2.9. If f is a D.C. function (that is, we can represent f as a difference of two continuous convex functions), then f is semismooth.

We will next show that these classes of functions verify condition (2.3) and therefore the generalized gradient of these functions can be characterized by (2.2), when they are Gateaux-differentiable on a dense subset in X.

Theorem 2.10. Let $f : X \to \mathbb{R}$ be a subregular or supregular function Gateaux-differentiable on a dense subset D in X; then the generalized gradient of f is characterized by formula (2.2).

Proof. We will prove that for all $x, d \in X$ such that the bilateral directional derivative $Df(x)(d)$ exists, we obtain the equality

$$Df(x)(d) = \lim_{\substack{y \to x \\ y \in D}} \langle \nabla f(y), d \rangle. \qquad (A)$$

If we define $\bar{f}(t) = f(x+td)$, we deduce that $\bar{f}$ is differentiable at 0, and by the sub or supregularity, jointly with the equality $\bar{f}^\circ(0;v) = f^\circ(x;vd)$ for any $v \in \mathbb{R}$, we conclude that

$$\{Df(x)(d)\} = \partial\bar{f}(0) = \langle \partial f(x), d \rangle$$

On the other hand, if $y_\alpha$ is a net in D converging to x, we know that $\nabla f(y_\alpha)$ has all its w*-accumulation points in $\partial f(x)$; therefore the net $\langle \nabla f(y_\alpha), d \rangle$ converges to $Df(x)(d)$. Since $y_\alpha$ is an arbitrary net converging to x, then equality (A) holds and by theorem 2.1 it follows that $\partial f(x)$ is characterized by (2.2)

□

Remark 2.11. In [2] it has been proved that the result of theorem 2.10 is also valid for the difference of two subregular functions, when both are Gateaux-differentiable on the same subset D in X. This class of nonsmooth functions which consists of the space spanned by the subregular (or the supregular) functions, has never been studied in the literature.

Theorem 2.12. If the function $f : X \to \mathbb{R}$ is semismooth and Gateaux-differentiable on a dense subset D in X, then the generalized gradient of f is characterized by formula (2.2).

Proof: As in theorem 2.10, we will demostrate that for all $x, d \in X$ such that the bilateral directional derivative $Df(x)(d)$ exists condition (2.3) is satisfied.

Since D is dense in X, it is clear that there exist both a net $t_\alpha$ in $\mathbb{R}_+$ converging to 0 and $d_\alpha$ in X converging to d such that $x+t_\alpha d_\alpha \in D$ for all $\alpha$. Since $\nabla f(x+t_\alpha d_\alpha) \in \partial f(x+t_\alpha d_\alpha)$ for all $\alpha$ and f is semismooth at x, we obtain

$$\lim_\alpha \langle \nabla f(x+t_\alpha d_\alpha), d \rangle = f'(x;d) = Df(x)(d),$$

which implies inequality (2.3). From theorem 2.1 we conclude the result.

□

Besides giving rise to new results, characterization (2.2) allows generalization of some other well-known ones occurring in finite dimensions. We will now characterize the subregular, supregular and semismooth functions in term of some continuity properties of the usual directional derivative (1.1).

Theorem 2.13. If the function $f : X \to \mathbb{R}$ is Gateaux-differentiable on a dense subset D in X and if $f'(x;d)$ exists for all $(x,d) \in X \times X$; then the following statements are equivalent:

(a) the function f is subregular (resp. supregular)

(b) i) $f'(\cdot;d)$ is u.s.c. (resp. $\ell$.s.c) on X for all $d \in X$
ii) $f'(\cdot;\cdot)$ is bounded in a neighbourhood of each point $(x,0) \in X \times X$

In the demonstration of this theorem we use the following lemma.

Lemma. Let $f : X \to \mathbb{R}$ be a function which is Gateaux-differentiable on a dense subset D in X and such that $f'(x;d)$ exists for all $(x,d) \in X \times X$. If f satisfies b) i), then it also satisfies (2.3).

Proof. For $x,d \in X$ such that $Df(x)(d)$ exists, we will show that

$$\lim_{\substack{y \to x \\ y \in D}} \langle \nabla f(y), d \rangle = Df(x)(d).$$

Since $f'(\cdot;d)$ is u.s.c. at x, we have the inequality

$$\limsup_{\substack{y \to x \\ y \in D}} \langle \nabla f(y), d \rangle \le Df(x)(d),$$

so that now we must only prove the converse inequality. Suppose that this is not true. Then there exists a net $y_\alpha$ converging to x with $y_\alpha \in D$ such that $Df(x)(d) = f'(x;d) > \lim_\alpha \langle \nabla f(y_\alpha), d \rangle + \varepsilon$ and from the $\ell$.s.c. of $-f'(\cdot;-d)$ at x we see that there exists a neighbourhood $V_x$ of x such that

$$-f'(y;-d) \ge \lim_\alpha \langle \nabla f(y_\alpha), d \rangle + \varepsilon \text{ for all } y \in V_x;$$

and since $y_\alpha$ converges to x, there exists $\alpha_0$ such that for $\alpha \ge \alpha_0$

$$-f'(y_\alpha;-d) = \langle \nabla f(y_\alpha), d \rangle \ge \lim_\alpha \langle \nabla f(y_\alpha), d \rangle + \varepsilon,$$

which is a contradiction.

The case $f'(\cdot;d)$ $\ell$.s.c. is analogous

□

Proof 2.13. (a) ⇒ (b). Trivial since $f^\circ(\cdot;d)$ is always u.s.c.

(b) ⇒ (a). From ii), for each $x \in X$ there exist an open neighbourhood $U \subset X$ of x, a neighbourhood $H \subset X$ of 0, and $\mu > 0$ such that $f'(x';d) \leq \mu$ for all $x' \in U$, $d \in H$. Hence if we choose a neighbourhood $V_o \subset X$ of x and a neighbourhood $H_o \subset H$ of 0 such that $V_o + H_o \subset U$, we obtain that $f'(x'+td;d) \leq \mu$ for all $x' \in V_o$, $d \in H_o$ and $t \in [0,1]$. Finally since a directionally differentiable function is the integral of its directional derivative, we deduce that f is Lipschitzian in a neighbourhood of x.

On the other hand, from the previous lemma and theorem 2.1, we have for all $x \in X$

$$\partial f(x) = \overline{co}\,\{w^*\text{-}\lim_{y\to x} \nabla f(y) / y \in D\},$$

which implies that $f^\circ(x;d) = \limsup_{\substack{y\to x\\ y\in D}} \langle \nabla f(y), d\rangle$ for all $x,d \in X$. Therefore using i) it follows that $f^\circ(x;d) \leq f'(x;d)$ (resp. $f_o(x;d) \geq f'(x;d)$). Since the converse inequality is always true, we conclude the demonstration.

Remark 2.14. If f is a locally Lipschitzian function on X then hypothesis ii) is superfluous.

An analogous theorem has been given by Rockafellar [13, theorem 2] in finite dimensions, in which hypothesis ii) is not necessary.

Theorem 2.15. Let $f : X \to \mathbb{R}$ be a locally Lipschitzian function and let D be a dense subset in X such that f is Gateaux-differentiable on D. Then the following are equivalent:

(a) f is semismooth.

(b) For all $x,d \in X$, the function $(t,h) \in \mathbb{R}_+ \times X \to f'(x+th;d)$ is continuous at $(0,d)$.

Proof. We shall demonstrate that (b) implies (a). We first note that a slight modification of the proof of theorem 2.12 shows that (b) implies (2.3) and, a fortiori, characterization (2.2) of the generalized gradient.

Let $x \in X$, let $(t_\alpha, h_\alpha) \in \mathbb{R}_+ \times X$ be a net converging to $(0,d)$ and $x^*_\alpha \in \partial f(x+t_\alpha h_\alpha)$. Given $\varepsilon > 0$, we will prove that there exists $\alpha_o$ such that

$$|\langle x^*_\alpha, d\rangle - f'(x;d)| < \varepsilon \text{ for all } \alpha \geq \alpha_o$$

By hypothesis there exist $\delta > 0$ and a neighbourhood $V_d$ of d such that

$$|\langle \nabla f(x+th), d\rangle - f'(x;d)| \leq \varepsilon/3 \text{ for all } t \in [0,\delta[,$$

where $h \in V_d$ and $x+th \in D$. Let $\alpha_o$ be such that for all $\alpha \geq \alpha_o$, $t_\alpha \in [0,\delta[$ and $h_\alpha \in V_d$. Since

$$x^*_\alpha \in \overline{\text{co}}\ \{w^*\text{-}\lim_{h \to h_\alpha} \nabla f(x+t_\alpha h) / x+t_\alpha h \in D\},$$

there exists $\bar{x}^*_\alpha \in \text{co}\ \{w^*\text{-}\lim_{h \to h_\alpha} \nabla f(x+t_\alpha h) / x + t_\alpha h \in D\}$, such that $|\langle x^*_\alpha - \bar{x}^*_\alpha, d\rangle| \leq \varepsilon/3$; therefore, for all $\alpha \geq \alpha_o$ there exists $x^{*k}_\alpha = w^*\text{-}\lim_\beta \nabla f(x+t_\alpha h^k_\beta)$ with $h^k_\beta \xrightarrow{\beta} h_\alpha$ and $x+t_\alpha h^k_\beta \in D$ for all $k \in K$ (finite subset of $\mathbb{N}$), and there also exists $\lambda^k_\alpha \geq 0$ with $\sum_{k \in K} \lambda^k_\alpha = 1$ such that $\bar{x}^*_\alpha = \sum_{k \in K} \lambda^k_\alpha x^{*k}_\alpha$.

Hence, for all $\alpha \geq \alpha_o$ there exists $\beta_\alpha$ such that $h^k_{\beta_\alpha} \in V_d$ and

$$|\langle x^{*k}_\alpha - \nabla f(x+t_\alpha h^k_{\beta_\alpha}), d\rangle| \leq \varepsilon/3.$$

Finally, by using the equality

$$\langle x^*_\alpha, d\rangle - f'(x;d) = \langle x^*_\alpha - \Sigma\lambda^k_\alpha x^{*k}_\alpha, d\rangle + \Sigma\lambda^k_\alpha \langle x^{*k}_\alpha - \nabla f(x+t_\alpha h^k_{\beta_\alpha}), d\rangle +$$
$$+ \Sigma\lambda^k_\alpha \langle \nabla f(x+t_\alpha h^k_{\beta_\alpha}), d\rangle - f'(x;d),$$

which is true for all $\alpha \geq \alpha_o$ we can conclude this part of the demonstration.

We now show that (a) implies (b). If f is semismooth at x, then for every net $(t_\alpha, h_\alpha) \in \mathbb{R}_+ \times X$ converging to $(0,d)$ one has in particular that the net

$$f'(x+t_\alpha h_\alpha; d) \in \langle \partial f(x+t_\alpha h_\alpha), d\rangle$$

converges to $f'(x;d)$, so the function $(t,h) \in \mathbb{R}_+ \times X \longrightarrow f'(x+th;d)$ is continuous at $(0,d)$.

□

Another interesting property of semismooth functions, which is not valid in general for locally Lipschitzian functions, is given by the following theorem.

Theorem 2.16. Let f,g be two locally Lipschitzian functions which are Gateaux-differentiable on a dense subset D of X. If they verify

i) $\nabla f(x) = \nabla g(x)$ for all $x \in D$

ii) f and g are semismooth,

then f-g is a constant function.

Proof. Since f-g is semismooth, it follows from theorem 2.12 that

$$\partial(f-g)(x) = \overline{co}\{w^*\text{-}\lim_{y \to x} \nabla(f-g)(y) / y \in D\} = \{0\} \quad \text{for all } x \in X.$$

On the other hand, from Lebourg's mean value theorem [8], one has that for all $y,z \in X$ there exists $u = y + t(z-y)$ with $t \in ]0,1[$ and $u^* \in \partial(f-g)(u)$ such that $(f-g)(y)-(f-g)(z) = \langle u^*, y-z \rangle = 0$, which implies the result.

□

Corollary. Let f,g be two locally Lipschitzian functions which are Gateaux-differentiable on a dense subset D of X. If they verify

i) $\partial f(x) \supset \partial g(x)$ for all $x \in X$

ii) $\partial f(x) = \{\nabla f(x)\}$ for all $x \in D$

iii) f is semismooth,

then f-g is a constant function.

We will have the same conclusion as in the above corollary if the hypothesis of semismoothness is replaced by the subregularity or supregularity of f [2] (in this case hypothesis ii) is superfluous). This result was given by Rockafellar [13] when X is finite dimensional.

## 3. Marginal functions.

An important representation of the locally Lipschitzian functions corresponds to the so-called marginal functions.

Definition 3.1. Let U be a topological space and let $g: X \times U \to \mathbb{R}$ be a function locally Lipschitzian on X uniformly in U, the function

$$f(x) = \sup_{u \in U} g(x,u)$$

is said to be the supmarginal function associated to g. We will assume that the solution set $M(x) = \{u \in U / f(x) = g(x,u)\}$ is nomempty.

The study of supmarginal functions consists of obtaining properties for the function f by using properties of the function g. An important class of supmarginal functions are the so-called lower-$C^k$ [13] [15] .

In what follows we will study the properties that must be satisfied by g for f to be subregular, or semismooth. We will also study the directional derivative and generalized gradient characterization for the function f in terms of the directional derivative and generalized gradient of $g(\cdot,u)$, which we will note $g_x'(x;u;d)$ and $\partial_x g(x,u)$ respectivelly.

We will begin by studying the directional derivative of a supmarginal function f. The most general result that we know in this sense, and which is given in lemma 3.4, uses the following definitions.

Definition 3.2. The lower Dini derivative of a function $f: X \to \mathbb{R}$ at a point $x \in X$ in the direction d is defined by the number

$$\underline{D}f(x;d) = \liminf_{t \to 0^+} \frac{f(x+td) - f(x)}{t} .$$

Definition 3.3. The multifunction M is said to be sequentially semi-continuous (s.s.c) at $x \in X$, if for every net $x_\alpha$ in X converging to x there exist $u \in M(x)$ and a net $u_\alpha \in M(x_\alpha)$ such that u is an accumulation point of $u_\alpha$ (i.e. there exists a subnet of $u_\alpha$ converging to u).

Lemma 3.4. Let $f : X \to \mathbb{R}$ be a supmarginal function. Let $x, d \in X$ be such that the multifunction $t \in \mathbb{R}_+ \to M(x+td)$ is s.s.c. at 0, and the function

$(t,u) \in \mathbb{R}_+ \times U \to \underline{D}_x\, g(x+td,u;d)$ is finite and u.s.c. on $\{0\} \times M(x)$. Then the directional derivative $f'(x;d)$ exists and can be characterized by

$$f'(x;d) = \sup \{g_x'(x,u;d) /\ u \in M(x)\}$$

Proof. Correa and Seeger [6]

Remark 3.5. If f is a supmarginal function, then M is s.s.c. at $x \in X$ if the function $g(x;\cdot)$ is u.s.c. on U and there exists a relatively compact selection $u(y) \in M(y)$ for y in some neighbourhood of x (in particular if U is a compact topological space). On account of the hypothesis of semicontinuity of the Dini derivative of g, in practice it is sufficient to verify the u.s.c. of the function $t \in \mathbb{R}_+ \to \underline{D}_x g(x+td,u;d)$ at 0, for each $u \in U$. In this sense we know that if the function $t \in \mathbb{R}_+ \to g(x+td,u)$ is semismooth, or subregular at 0, then the function $t \in \mathbb{R}_+ \to \underline{D}_x g(x+td,u;d)$ is continuous or u.s.c. at 0 respectively.

We now give a characterization of the directional derivative and generalized gradient of a supmarginal function, when the functions $g(\cdot,u)$ are subregular. This result generalizes a similar one given by Clarke [5, theorem 2.8.2.] .

Theorem 3.6. If $f: X \to \mathbb{R}$ is a supmarginal function, the multifunction M is s.s.c. at x, the function $g(\cdot,u)$ is subregular at x for each $u \in M(x)$, and the multifunction $\partial_x g(\cdot,\cdot)$ is u.s.c. on $\{x\} \times M(x)$, then f is subregular at x and its directional derivative and generalized gradient can be characterized by the formulas

$$f'(x;d) = \sup \{g_x'(x,u;d) / u \in M(x)\}$$

$$\partial f(x) = \overline{co}\ \{\partial_x g(x,u)\ /\ u \in M(x)\}$$

Proof. 1) For all $u \in M(x)$, $d \in X, x^* \in \partial_x g(x,u)$, and from the fact that $g(\cdot,u)$ is subregular, we have

$$< x^*,d > \le g_x^\circ(x,u;d) = g_x'(x,u;d) = \lim_{t \to 0^+} t^{-1}[g(x+td,u) - g(x,u)]$$

$$\le \limsup_{t \to 0^+} t^{-1}\ [f(x+td) - f(x)] \le f^\circ(x;d),$$

that is $\partial f(x) \supset \overline{co}\ \{\partial_x g(x,u)\ /\ u \in M(x)\}$

2) Let $x_\alpha \to x$, $t_\alpha \to 0^+$ so that $f^\circ(x;d) = \lim_\alpha \dfrac{f(x_\alpha + t_\alpha d) - f(x_\alpha)}{t_\alpha}$ ;

on the other hand, for all $v_\alpha \in M(x_\alpha + t_\alpha d)$ there exists $x^*_\alpha$ such that

$$\frac{f(x_\alpha + t_\alpha d) - f(x_\alpha)}{t_\alpha} \le \frac{g(x_\alpha + t_\alpha d, v_\alpha) - g(x_\alpha , v_\alpha)}{t_\alpha} = \langle x^*_\alpha, d \rangle ,$$

where $x^*_\alpha \in \partial_x g(y_\alpha, v_\alpha)$ and $y_\alpha \in ]x_\alpha, x_\alpha + t_\alpha d[$ (mean value theorem of Lebourg). Since M is s.s.c. at x, there exists an accumulation point $v \in M(x)$ of a net $v_\alpha$. Since $y_\alpha$ converges to x and $\partial_x g(\cdot,\cdot)$ is u.s.c. on $\{x\} \times M(x)$, there exists a weak*-accumulation point $x^* \in \partial_x g(x,v)$ of the net $x^*_\alpha$ . So, for all $d \in X$ there exists $v \in M(x)$ such that

$$f^\circ(x;d) \le \langle x^*, d \rangle \le g^\circ(x,v;d) \tag{A}$$

which implies $\partial f(x) \subset \overline{co} \{\partial_x g(x,u) \ / \ u \in M(x)\}$.

3) The hypotheses of lemma 3.4 are implied by those of this theorem. In fact, only the u.s.c. of the function $(t,u) \in \mathbb{R}_+ \times U \to \underline{D}g(x+td,u;d)$ on $\{0\} \times M(x)$ is not trivial; however, this fact follows easily from the u.s.c. of $g^\circ_x(\cdot,\cdot;d)$ and the inequality $\underline{D}_x g(x+td,u;d) \le g^\circ_x(x+td,u;d)$ (since one has the equality in $t = 0$). Therefore the characterization of $f'(x;d)$ holds true, and using (A) we have

$$\max \{g'_x(x,u;d) \ / \ u \in M(x)\} = f'(x;d) \le f^\circ(x;d) \le g^\circ_x(x,v;d)$$

$$= g'_x(x,v;d) \le \max \{g'_x(x,u;d) \ / \ u \in M(x)\},$$

which implies the subregularity of f at x.

□

We now give sufficient conditions for a supmarginal function to be semismooth and we characterize its directional derivative. We will assume that f and $g(\cdot,u)$ are Gateaux-differentiable on a dense subset of X.

Definition 3.7. A function $g : X \times U \to \mathbb{R}$ is said to be semismooth on X continuously on $M \subset U$ provided that, for all x, $d \in X$ the function $(t,h,u) \in \mathbb{R}_+ \times X \times M \to g'_x(x+th,u;d)$ is continuous on $\{0\} \times \{d\} \times M$.

Theorem 3.8. Let f be a supmarginal function such that the topological space U is compact and $g(x,\cdot)$ is u.s.c. on U for all $x \in X$. If g is semismooth on X continuously on U then f is semismooth, and one has

$$f'(x;d) = \sup \{g'_x(x,u;d) \ / \ u \in M(x)\}$$

Proof. The hypotheses of lemma 3.4 are easily verified by using theorem 2.15. to characterize the semismoothness of the functions $g(\cdot,u)$ in terms of continuity of its directional derivatives (see remark 3.5 for the s.s.c. of M). Then we obtain the existence of $f'(x;d)$ and its characterization $f'(x;d) = \sup\{g'_x(x,u;d)/ u\in M(x)\}$.

By using again theorem 2.15, we see that in order to prove the semi - smoothness of f it is sufficient to show that for every net $(t_\alpha,h_\alpha)\in \mathbb{R}_+ \times X$ converging to $(0,d)$ one has $\lim_\alpha f'(x_\alpha;d) = f'(x;d)$ where $x_\alpha = x + t_\alpha h_\alpha$.

Since $f'(x_\alpha;d) = \sup\{g'_x(x_\alpha,u;d)/u\in M(x_\alpha)\}$ and from the fact that $M(x_\alpha)\subset U$ is compact, there exists $u_\alpha\in M(x_\alpha)$ such that $f'(x_\alpha;d)=g'_x(x_\alpha,u_\alpha;d)$, and convergent subnets of $(u_\alpha)$. Let $(v_\beta)$ be a subnet of $(u_\alpha)$ converging to $\bar{u}$, and let $(y_\beta)$, $(k_\beta)$ and $(r_\beta)$ be the corresponding subnets of $(x_\alpha)$, $(h_\alpha)$ and $(t_\alpha)$ respectively. From the definition of supmarginal functions and the u.s.c. of $g(y,\cdot)$ it follows that the multifunction $(t,h)\to M(x+th)$ is closed at $\{0\}\times\{d\}$, and, therefore, $\bar{u}\in M(x)$. Then, in order to prove this theorem it is sufficient to show that $\bar{u}\in V(x)=\{u\in M(x)/f'(x;d)=g'(x,u;d)\}$, i.e. the multifunction $(t,h)\to V(x+th)$ is closed at $\{0\}\times\{d\}$. Let us suppose that $\bar{u}\notin V(x)$; then there exist $\varepsilon > 0$ and $u_o\in V(x)$ such that

$$g'_x(x,\bar{u};d) \le g'_x(x,u_o;d) - \varepsilon \qquad \text{(A)}$$

On the other hand, from the fact that $g'_x(x+th,u;\cdot)$ is Lipschitzian uniformly on $(t,h,u)$ and the function $(t,h,u)\to g'_x(x+th,u;d)$ is continuous on $\{0\}\times\{d\}\times M(x)$, it follows that the function $(t,h,u)\to g'_x(x+th,u;h)$ is continuous on $\{0\}\times\{d\}\times M(x)$. Hence from (A) it follows that there exists $\beta_1$ such that

$$g'_x(x+tk_\beta,v_\beta;k_\beta) \le g'_x(x+tk_\beta,u_o;k_\beta) - \varepsilon/2$$

for all $\beta > \beta_1$ and $t\in[0,r_\beta]$. Since a directionally differentiable function is the integral of its directional derivative we deduce from the last inequality that $g(y_\beta,v_\beta)-g(x,v_\beta) \le g(y_\beta,u_o)-g(x,u_o)-\varepsilon r_\beta/2$ which means that $f(y_\beta)+f(x)\le g(y_\beta,u_o)+g(x,v_\beta)-\varepsilon r_\beta/2$ which is a contradiction. Therefore, for any convergent subnet $(v_\beta)$ of $(u_\alpha)$ the net $(f'(y_\beta;d))$ has the same limit $f'(x;d)$. Thus since U is compact we have $f'(x_\alpha;d) \underset{\alpha}{\to} f'(x;d)$. □

The above theorem shows that the lower-$C^1$ functions are semismooth.

REFERENCES

[ 1 ] Aronszajn,N., Differentiability of Lipschitzian mappings between Banach spaces, Studia Math. 57 (1976), 147-190.

[ 2 ] Bustos,M., Correa R. and Tapia L., On the generalized gradient of some classes of nonsmooth functions, to appear in M.A.N. 1985.

[ 3 ] Christensen, J.P., Topology and Borel structure, Math. Studies nº 10, North-Holland Amsterdam, 1974.

[ 4 ] Clarke, F.H., Generalized gradients and applications, Trans. Amer. Math. Soc. 205(1975), 247-262.

[ 5 ] Clarke,F.H., Optimization and Nonsmooth Analysis, Wiley - Interscience, Pub.,1983.

[ 6 ] Correa,R. and Seeger,A., Directional derivative of a minimax function, Nonlinear Analysis, (1985), 13-22.

[ 7 ] Jofré,A., Regular and semismooth functions in nonsmooth analysis, Thesis of Engineer, Universidad de Chile, 1984.

[ 8 ] Lebourg,G., Generic differentiability of Lipschitzian functions, Trans. Amer. Math. Soc. 256(1979), 125-144.

[ 9 ] Mifflin, R., Semismooth and semiconvex functions in constrained optimization SIAM on Control and Opt. 15(1977), 959-972.

[10 ] Mignot,F., Un theoreme de differentiabilité; application aux proyections et au controle dans les inequations. Sém. sur les equations aux derivées partielles, Université-Paris VI.(1973-1974).

[ 11 ] Moreau,J.J., Funcionelles convexes. Sém. sur les equations aux derivées partielles. College de France (1966).

[ 12 ] Rockafeller, R.T., Convex Analysis, Princeton Mathematics.Ser. Vol. 28, Princeton Univ. Press. (1970).

[ 13 ] Rockafellar, R.T., Favorable classes of Lipschitz continuous functions in subgradient optimization, in Nondifferentiable Optimization, E.Nurminski, Ed., Pergamon Press,New York(1982).

[ 14 ] Shu-Chung,S., Remarques sur le gradient généralisé, C.R.A.S. Paris 291,(1980),443-446.

[ 15 ] Spingarn,J.E., Submonotone subdifferentials of Lipschitz functions, Trans. Am. Math. Soc. 264,(1981),77-89.

[ 16 ] Thibault,L., On generalized differentials and subdifferentials of Lipschitz vector-valued functions, Nonlinear Analysis 6, (1982),1037-1053.

[ 17 ] Zivkov,N., Generic Gateaux differentiability of locally Lipschitzian functions. Constructiv Function Theory'81, Sofia,(1983), 590-594.

# MODELLING ERRORS IN TIME SERIES AND K-STEP PREDICTIONS.

G.E. del Pino and P. Marshall
Departamento de Probabilidad
y Estadística. Universidad
Católica de Chile. Santiago,
Chile.

ABSTRACT.

ABSTRACT.

The effect of specification error for ARIMA models both on the actual k-step prediction error variances and on the validity of the usual formulae for these variances is studied theoretically and empirically.

## 1.- INTRODUCTION

When fitting ARIMA models to time series, it must be kept in mind that the model specified will only be an approximation to the true one. In this paper we analyze the effect of this misspecification on the variances of k-step prediction errors. For this purpose we define the concepts of optimal, effective and apparent variances. A comparison of effective with the optimal variance will give an idea of the loss of prediction efficiency, while a comparison with the apparent variance will give and idea of how seriously the usual theoretical formulae for k-step error variances will be invalidated.

In chapter 2 the basic concepts are defined and fundamental formulare are derived. In chapter 3 the important case where the specified model is autoregressive is examined. The role of the method of moments and partial autocorrelations are pointed out. In chapter 3 there is a detailed theoretical examination of some stationary models, assuming that the criterion for fitting them is to minimize the one step prediction error variance. In chapter 5 the case where the true model is AR(1) while the specified model is ARIMA (0,1,1) is examined. Finally, in chapter 6 the results of some simulations are discussed. These are important to evaluate the relevance of the theoretical analyses in the previous chapters.

## 2.- BASIC CONCEPTS AND FORMULAE.

Let $X_t$ be a stationary ARIMA model with MA($\infty$) representation

$$X_t = \Psi(B)a_t \tag{2.1}$$

and let $e_k$ be the optimal prediction error of $X_{t+k}$ given $X_s$, $s \leq t$. Assume that a wrong ARIMA model is fitted, whose MA($\infty$) representation is given by

$$X_t = \Psi'(B)u_t \tag{2.2}$$

This model imply a different prediction of $X_{t+k}$, given $X_s$, $s \leq t$ and a new prediction error $e_k'$ .

The variances $V_k^O = V(e_k)$ and $V_k^E = V(e_k')$ will be called *optimal* and *effective* variances respectively. The variance of $e_k'$ calculated as if $u_t$ were white noise (i.e. as if model (2.2) were correct) will be denoted by $V_k^A$ and will be called *apparent* variance. We will be interested in the relationships between $V_k^O$, $V_k^E$ and $V_k^A$ as well as between their asymptotic values $V^O$, $V^E$ and $V^A$ when $k$ tends to infinity. (These limits will certainly exist for stationary models). The use of a wrong model will increase the prediction error variance. The quantity

$$INEF(k) = (V_k^E/V_k^O) - 1 \tag{2.3}$$

will be called inefficiency of order $k$. The limit of INEF($k$) as $k$ tends to $\infty$ will be denoted by INEF($\infty$) and gives an idea of the precision lost in long term predictions.

We will need the following notation:

For $$L(B) = \sum_0^{\infty} L_i B^i$$

let $$L_k(B) = \sum_0^{k-1} L_i B^i$$

$$F(L,k) = \sum_0^{k-1} L_i^2$$

and $$F(L) = F(L,\infty)$$

For convenience we will also assume $\sigma_a^2 = 1$. Let

$$\Psi''(B) = (\Psi'(B))^{-1}\Psi(B), \tag{2.4}$$

It is then easy to prove

$$V_k^O = F(\Psi,k) \tag{2.5}$$

$$V_k^E = F(\Psi_k' \Psi'') \tag{2.6}$$

and $$V_k^A = F(\Psi'')F(\Psi',k) \tag{2.7}$$

From (2.4), (2.5) and (2.6) it follows that $V^E$ and $V^O$ coincide if they are finite, so that INEF($\infty$) $= 0$.

As far as $V_k^A$, the following theorem may be proved by expressing variances in terms of spectral densities.

<u>Theorem 2.1.</u> *Let $U$ be uniform on $[-\pi,\pi]$, let $f(u) = |\Psi_k'(e^{-iu})|^2$, $g(u) = |\Psi''(e^{-iu})|^2$ and $\gamma = Cov(f(U),g(U))$. Then $V_k^E \gtreqless V_k^A$ if and only if $\gamma \gtreqless 0$.*

Corollary 2.1. *Let f and g defined in Theorem 2.1 be monotonic. If, f and g are both increasing or both decreasing* $V_k^E \geq V_k^A$, otherwise $V_k^E \leq V_k^A$ .

## 3.- FITTED AUTOREGRESSIVE MODEL

A particularly interesting situation occurs when the fitted model is AR($p$), i.e.

$$X_t = \pi_1 X_{t-1} + \dots + \pi_p X_{t-p} + u_t \tag{3.1}$$

Let $\pi = (\pi_1,\dots,\pi_p)'$ and consider the problem

$$\min_{\pi} Var(X_t - \pi_1 X_{t-1} - \dots - \pi_p X_{t-p}) \tag{3.2}$$

But (3.2) defines the best linear predictor of $X_t$ given $X_{t-1},\dots,X_{t-p}$. It is known that the solution of (3.2) coincides with the solution $\hat{\pi}$ of

$$A\pi = b, \tag{3.3}$$

where

$$A_{ij} = \rho_{|i-j|} \qquad i,j = 1,\dots,p$$
$$b_i = \rho_i \qquad i = 1,\dots,p.$$

The system (3.3) corresponds to the so called Yule Walker equations. Furthermore the minimum in (3.2) is given by

$$Var(X_t)(1 - \sum_{i=1}^{p} \hat{\pi}_i \rho_i) \tag{3.4}$$

Consider now the $k$-step predictions. A linear predictor of $X_{t+k}$ given $X_1,\dots,X_t$ corresponding to an AR($p$) model, takes the form

$$\sum_{1}^{p} a_i X_{t+1-i} ,$$

where $a_i = a_i(\pi_1,\dots,\pi_p)$, $i = 1,\dots,p$. On the other hand the best linear predictor of the form

$$\sum_1^p c_i X_{t+1-i}$$

i.e. the one that minimizes

$$Var(X_{t+k} - \sum_1^p c_i X_{t+1-i})^2 \tag{3.5}$$

is obtained by choosing $c_i = \hat{c}_i$, $i = 1,\ldots,p$, where $\hat{c} = (\hat{c}_1,\ldots,\hat{c}_p)'$ is the solution of

$$Ac = b^{(k)}, \tag{3.6}$$

with $\qquad b_i^{(k)} = \rho_{i+k-1} \qquad i = 1,\ldots,p$

and the minimum value of (3.5) is

$$Var\ X_t(1 - \sum_1^p \hat{c}_i \rho_{i+k-1}) \tag{3.7}$$

If we denote the values of $\pi_1,\ldots,\pi_p$ minimizing the $k$-step error variance by $\hat{\pi}_1^{(k)},\ldots,\hat{\pi}_p^{(k)}$ and if $\hat{\pi}^{(k)} = (\hat{\pi}_1^{(k)},\ldots,\hat{\pi}_p^{(k)})'$, then $\hat{\pi}^{(k)}$ is the solution of

$$a_i(\hat{\pi}^{(k)}) = \hat{c}_i \qquad i = 1,\ldots,p \tag{3.8}$$

although this may not be unique.

Let now $X_t$ be an stationary series and let

$$\sigma_k^2 = \min_{a_1,\ldots,a_k} Var(X_t - a_1X_{t-1} - \ldots - a_kX_{t-k})^2, \tag{3.9}$$

It is clear that $\sigma_k^2$ is non increasing in $k$. More precisely, if $c_k$ denotes the $k$-$th$ partial autocorrelation, then

$$\sigma_k^2 = \sigma_{k-1}^2(1 - c_k^2) \qquad k \geq 1, \tag{3.10}$$

with $\sigma_0^2 = VarX_t$.

Fitted model AR($p$-1). An application of (3.10) shows that

$$INEF(1) = \phi_p^2/(1-\phi_p^2) \tag{3.11}$$

It is also possible in this case to give simple expressions for

$\hat{\pi}_1, \ldots, \hat{\pi}_{p-1}$:

$$\hat{\pi}_k = (1-\phi_p^2)^{-1}(\phi_k + \phi_p\phi_{p-k}) \qquad k = 1,\ldots,p-1 \qquad (3.12)$$

## 4.- THEORETICAL ANALYSIS OF SOME MODELS.

In this section we examine analitically the effect of fitting wrong models for some special cases, in which it is possible to obtain simple expressions and cualitative information. The cases considered are

$A : X_t = (1-\theta B)a_t \qquad vs. \qquad (1-\pi B)X_t = u_t$

$B : (1-\phi B)X_t = (1-\theta B)a_t \qquad vs. \qquad (1-\pi B)X_t = u_t$

$C : X_t = (1-\theta_1 B - \theta_2 B^2)a_t \qquad vs. \qquad X_t = (1-\Psi B)u_t$

$D : (1- B)X_t = a_t \qquad vs. \qquad X_t = (1-\Psi B)u_t$

where the model in the left is the true one and the other is the fitted one.

The main results are summarized in Table 4.1.

TABLE 4.1

| | A | B | C | D |
|---|---|---|---|---|
| Fitted parameter | $\hat{\pi} = -\theta/(1+\theta^2)$ | $\hat{\pi} = (\phi-\theta)(1-\phi\theta)/(1+\theta^2-2\theta\phi)$ | $\hat{\Psi} = \theta_1/(1-\theta_2)$ | $\hat{\Psi} \simeq -\phi+\phi^3/(1+3\phi^2)$ |
| INEF (1) | $\theta^4/(1+\theta^2)$ | $\theta^2[1-(1-\phi^2)/(1+\theta^2-2\phi\theta)]$ | $\theta_2^2$ | $<\phi^4/((1-\phi^4)$ |
| Least favorable case | $\theta = \pm 1$ | $(\phi,\theta) = (1,-1),(-1,1)$ | $\theta_2 = \pm 1$ | $\phi = 1$ |
| Maximum INEF (1) | $1/2$ | $1$ | $1$ | $\infty$ |

For case A it is seen that if $|\theta| < 0.5(0.1)$, then INEF(1) does not exceed $0.05$ $(0.001)$. This suggests on the one hand that it should be difficult to discriminate between the two models, and on the other that the possible error will not be serious.

It may additionally be proved that

$$V_k^E = (1+\theta^2)(1 + \hat{\pi}^{2k})$$

and

$$V_k^A = (1+\theta^2)(1 - \hat{\pi}^{2k}),$$

so that the apparent variance underestimates the effective variance. Modelling errors are again negligible for large $k$.

The analysis of case B is more complex, in the sense that there are two parameters $\phi,\theta$. For a fixed value of $\theta$, INEF(1) attains a maximum of $\theta^2$ for $\phi = \pm 1$ while for a fixed value of $0 \leq \phi < 1$ INEF(1) shows three relative maxima : at $\theta = 1,-1,a$ with respective value of $(1-\phi)/2$, $(1+\phi)/2$ and $(\phi-2a)(\phi-a)$, where $a$ is the solution of

$$x^3 = (1-\phi^2)(\phi-2x)$$

on the interval $(0,\phi/2)$.

It would be interesting to generalize case C to the problem of fitting MA($q-1$) when the true model is MA($q$). Numerical computations and the results obtained for $q = 2$ suggest that

$$INEF(1) = \theta_q^2$$

will hold in general, but at the moment of writting this remains a conjecture.

In case D it is not possible to obtain closed expressions for $\hat{\theta}$.

It may be proved, that $\hat{\theta}$ is the minimizer of

$$R(\theta) = \frac{(1-\phi\theta)(\phi+\theta)^2}{1+\theta\phi} + \phi^2\theta^2 \tag{4.1}$$

We remark that $\hat{\theta}(-\phi) = -\hat{\theta}(\phi)$. This minimization problem may be solved iteratively. For small $\phi$, suitable initial values are

$$\theta_1 = -\phi$$

and

$$\theta_2 = [(1-4\phi^2)^{1/2} - 1]/\phi ,$$

this last one corresponding to the method of moments.

An application of one step Newton-Raphson with starting point $\theta_1$ to the minimization of (4.1) gives

$$\theta_3 = -\phi + \frac{\phi^3(1-\phi^2)}{1+2\phi^2-\phi^4}$$

Finally, if $Var(u_t)$ is computed approximating $u_t$ by an MA(2) model, the problem reduces to minimizing

$$(\phi+\theta)^2 + ((\phi+\theta)^2 - \phi\theta)$$

and an application of one step Newton Raphson approximates $\hat{\theta}$ by

$$\theta_4 = -\phi + \phi^3/(1+3\phi^2)$$

It is easily shown that for $\phi > 0$, $\theta_2 < \theta_1 < \theta_3 < \theta_4$. These approximations to $\hat{\theta}$ should be better for small $\phi$. For $\phi = 0.4$, $\hat{\theta} = -.3564$ while $\theta_2 = -.5000$, $\theta_1 = -.4000$, $\theta_3 = -.3574$ and $\theta_4 = -.3568$.

## 5.- A NON STATIONARY EXAMPLE.

Consider now the case where the true model is

$$(1-\phi B)(X_t-\mu) = a_t$$

but

$$(1-B)X_t = (1-\theta B)u_t$$

is fitted. It may be shown that

$$Var(u_t) = 2[(1+\theta)(1+\phi)(1-\theta\phi)]^{-1} \tag{5.1}$$

For the sake of simplicity, and also because is the most important case in practice, we consider only $\phi > 0$. The minimizer $\hat{\theta}$ of (5.1) is

$$\begin{aligned} \hat{\theta} &= 1 \quad \text{if} \quad \phi \le 1/3 \\ &= (1-\phi)/2\phi \quad \text{if } \phi \ge 1/3 \end{aligned} \tag{5.2}$$

Substituting (5.2) into (5.1) it follows that

$$\begin{aligned} (Var\ u_t)(\hat{\theta}) &= (1-\phi^2)^{-1} & \phi \le 1/3 \\ &= 8\phi(1+\phi)^{-3} & \phi \ge 1/3 \end{aligned}$$

and from this, the maximum value of INEF(1) is $5/27$, which is attained at $\phi = .5$. For $\phi = 0$ and $\phi = 1$, INEF(1) is zero as expected.

For $k$-step predictions, a rather lengthy computation shows that

$$V_k^E = \frac{1}{1-\phi^2}[1 + \frac{(1-\theta)}{(1-\theta\phi)}(\frac{1+\theta\phi}{1+\theta} - 2\phi^k)] \tag{5.3}$$

which reduces to (5.1) for $k = 1$.

It is also easily proved that

$$V_k^O = (1-\phi^{2k})/(1-\phi^2) \tag{5.4}$$

and it then follows that, unlike the stationary case, $V^O \neq V^E$.

If $1 > \phi \geq 1/3$ and $\theta = \hat{\theta}$ given by (5.2), then

$$\text{INEF}(\infty) = \frac{(3\phi-1)(3-\phi)}{(1+\phi)^2} \tag{5.5}$$

The limit value of (5.5) as $\phi$ tends to $1$ is $1$, which seems to contradict the intuitive fact that for $\phi = 1$, $\text{INEF}(\infty) = 0$. This contradiction is only apparent since it may be checked that for $\phi$ near $1$

$$\text{INEF}(k) = \varepsilon^2/4k + O(\varepsilon^3)$$

with $\varepsilon = 1-\phi$, and this gives the right limit.

## 6.- SOME NUMERICAL AND EMPIRICAL RESULTS.

With the purpose of evaluating the effects of fitting wrong models, we present some numerical and empirical experiences. The cases analyzed were

| Case | True model | Fitted model |
|---|---|---|
| A | $(1-.5B)X_t = a_t$ | $X_t = (1-\psi B)u_t$ |
| B | $X_t = (1-.5B)a_t$ | $(1-\pi B)X_t = u_t$ |
| C | $(1-.6B-.2B^2)X_t = a_t$ | $(1-\pi B)X_t = u_t$ |
| D | $X_t = (1-.6B-.2B^2)a_t$ | $X_t = (1-\psi B)u_t$ |
| E | $(1-.8B)X_t = (1-.4B)a_t$ | $(1-\pi B)X_t = u_t$ |
| F | $(1-.5B)X_t = a_t$ | $(1-B)X_t = (1-\psi B)u_t$ |

### Numerical study.

The value of the parameter minimizing the effective one step variance was computed using the formulae obtained in the previous sections. Then INEF($k$) was computed theoretically using a general computer program specially developed for this purpose. The results of these computations are presented in Table 6.1. Perhaps the main aspects here are that for the stationary cases (A-E) the INEF($k$) values are rather small and that the maximum values are not necessarily attained at $k = 1$ (Remember that $\lim_{k\to\infty}$ INEF($k$) = 0). In case $D$ ($MA(2)$ vs. $MA(1)$) INEF($k$) attains its limiting value at $k = 3$. In case $F$, as expected from (5.5), INEF($k$) increases to a limiting value of 5/9.

TABLE 6.1 THEORETICAL VALUES OF INEF($k$) (in %)

| | CASE | | | | | |
|---|---|---|---|---|---|---|
| K | A | B | C | D | E | F |
| 1 | 5.7 | 5.0 | 4.2 | 4.0 | 4.9 | 18.5 |
| 2 | 6.7 | 2.6 | 2.4 | 2.9 | 5.3 | 30.4 |
| 3 | 1.6 | 0.4 | 2.8 | 0.0 | 5.8 | 41.1 |
| 4 | 0.4 | 0.1 | 2.9 | 0.0 | 5.0 | 47.8 |
| 5 | 0.1 | 0.0 | 2.7 | 0.0 | 3.8 | 51.5 |

Simulation study.

For each of the cases considered 5 realizations of length 200 were simulated according to the true model. Then the parameters of the model were estimated by the statistical package IDA(which uses back forecasting).

The effective variance $VE(k)$ was estimated as the mean square $\hat{V}E(k)$ of the observed $k$-step prediction errors, while the estimator $VA(k)$ of the apparent variance was obtained replacing the parameters by the estimated parameters in the theoretical formulae. Finally for several values of $k$, $A/E(k)$ defined by

$$A/E(k) = 100[\hat{V}A(k)/\hat{V}E(k) - 1] \tag{6.1}$$

was evaluated.

For the sake of space we only include the results corresponding to cases $E$ and $F$, which are shown on Tables 6.2 and 6.3 respectively. Nevertheless the general comments below are applicable to all the cases. Apart from $A/E$, the theoretical equivalent computed from the known parameter values of the true and fitted model is shown in the first column of these tables.

Despite the relative closeness of the estimators for the different realizations and despite the length of the series, a great dispersion is observed among the values of $A/E$. There are also great discrepan - cies between them and the theoretical values. However the general behavior of $A/E$ as a function of $k$ was very stable for the different realizations.

As it was expected theoretically, in the nonstationary case $A/E$

without limit.

The rather large values of $A/E$, even when fitting models close to the true one, give some hope of using this quantity as a tool for detecting lack of fit.

TABLE 6.2 SIMULATION STUDY OF $A/E$ FOR CASE E

| | Theoretical | Empirical | | | | |
|---|---|---|---|---|---|---|
| | $\pi$ = .52 | $\pi$ = .51 | $\pi$ = .63 | $\pi$ = .54 | $\pi$ = .50 | $\pi$ = .36 |
| $k$ | $A/E$ | $A/E$ | $A/E$ | $A/E$ | $A/E$ | $A/E$ |
| 2 | 9.4 | 6.0 | 15.7 | 16.0 | 13.7 | 5.5 |
| 3 | 5.9 | 5.8 | 9.5 | 7.4 | 6.0 | 2.6 |
| 4 | 3.0 | 3.4 | 5.3 | 4.9 | 3.7 | 1.6 |
| 5 | 1.4 | 2.4 | 1.8 | 1.4 | 3.8 | 0.7 |
| 6 | 0.6 | 1.9 | -0.1 | 0.0 | 2.5 | 0.0 |
| 7 | 0.3 | 1.3 | 0.3 | -0.8 | 3.1 | -0.5 |
| 8 | 0.1 | 1.2 | -0.3 | -1.5 | 2.8 | -1.0 |
| 9 | 0.0 | 1.4 | 0.3 | -1.8 | 2.9 | -1.5 |
| 10 | 0.0 | 3.9 | -0.1 | -2.3 | 2.3 | -2.0 |
| $\infty$ | 0.0 | 2.6 | 0.9 | 0.9 | 0.0 | 0.3 |

TABLE 6.3 SIMULATION STUDY OF $A/E$ FOR CASE F

| | Theoretical | Empirical | | | | |
|---|---|---|---|---|---|---|
| | $\psi$ = .50 | $\psi$ = .61 | $\psi$ = .54 | $\psi$ = .76 | $\psi$ = .68 | $\psi$ = .55 |
| $k$ | $A/E$ | $A/E$ | $A/E$ | $A/E$ | $A/E$ | $A/E$ |
| 2 | -9.1 | -11.5 | -9.3 | -11.4 | -11.8 | -7.3 |
| 3 | -4.0 | -7.4 | -3.6 | -9.2 | -8.5 | -5.9 |
| 4 | 5.7 | -3.0 | 1.7 | -4.3 | -5.2 | -3.3 |
| 5 | 17.4 | 6.1 | 10.3 | 2.3 | 3.3 | 7.0 |
| 6 | 30.3 | 17.3 | 21.2 | 7.3 | 14.1 | 21.2 |
| 7 | 43.8 | 31.7 | 35.2 | 8.4 | 15.5 | 34.4 |
| 8 | 57.7 | 42.0 | 39.3 | 9.7 | 17.8 | 46.4 |
| 9 | 71.7 | 50.8 | 45.5 | 9.0 | 22.4 | 64.7 |
| 10 | 85.9 | 60.4 | 46.2 | 8.0 | 29.2 | 84.7 |

# REFERENCES

Box, G.E.P. and Jenkins, G. (1976). Time Series Analysis, Forecasting and Control. Revised Edition. San Francisco, Holden Day.

Findley, D.F. (1983). On using a different time series model for each forecast lead. Bureau of the Census, SRD/RR-83/06.

Findley, D.F. (1983). On ambiguities associated with fitting ARMA models to time series. Presented in NSF-NBER - University of Chicago Seminar on Time Series, October 7-8, 1983.

# THE SINGLE DEPOT VEHICLE SCHEDULING PROBLEM

Jacques A. Ferland
Centre de recherche sur les transports and
Département d'informatique et de recherche opérationnelle
Université de Montréal
P.O.Box 6128, Station A,
Montréal (Québec) CANADA H3C 3J7

## 1. INTRODUCTION

The vehicle scheduling problem can be specified in terms of a set of tasks to be executed with a fleet of vehicles. Each task i includes time and space (location) characteristics specified as follows :

- . a start location $SL_i$
- . an end location $EL_i$
- . a start time $ST_i$
- . a time duration $D_i$
- . an end time $ET_i = ST_i + D_i$.

To execute task i a vehicle must be at location $SL_i$ at time $ST_i$. Then at time $ET_i$ the vehicle is located at $EL_i$ and becomes available for other tasks. The problem is to partition the set of tasks into sequences of tasks, with each sequence corresponding to the schedule of a vehicle. Hence no two tasks of a sequence can take place simultaneously. Furthermore task j can be executed after task i in the sequence if, taking into account the time $t_{ij}$ required for a vehicle to go from $EL_i$ to $SL_j$, the vehicle can be made available at time $ST_j$ at $SL_j$; i.e.

$$ET_i + t_{ij} \leq SL_j.$$

The fleet of vehicles available can be characterized by its size (the number of vehicles), by the locations of the depots where the vehicles start and end their schedules, and by the types of vehicles, etc. In this paper we are concerned with the problem in which there is a single depot and all vehicles are of the same type.

The objective function to be minimized can be specified in terms of operating costs, or in terms of investment costs, or in terms of both. For instance, the objective might be to reduce the fleet size to a minimum.

The vehicle scheduling problem is presented and analyzed in Bodin et al. (1983) where procedures for handling different variants are summarized. The purpose of this paper is to complete this survey for the variant with a single depot. In particular, many authors have recently addressed the problem of vehicle scheduling with time window constraints where the start time $ST_i$ of each task i must lie within some time interval. The purpose of this paper is to summarize the procedures rather than to analyze and comment on their relative efficiency.

In Section 2 we analyze the case where the start time $ST_i$ of each task i is fixed. Then in Section 3 we turn to the vehicle scheduling problem with time window constraints. Heuristic and exact (branch-and-bound) approaches are reviewed in Sections 4 and 5, respectively.

## 2. FIXED START TIME

A mathematical formulation is derived for this problem using a precedence graph of the tasks, (N,A), characterized as follows :

i) With each task i associate a node i in N. Two nodes, s and t, are associated with the depot to indicate where the vehicles start and finish.

ii) An arc (i,j) exists (i.e., $(i,j) \varepsilon A$, or task j can follow task i) if

$ET_i + t_{ij} \leq ST_j$.

Furthermore, for each task i, introduce arcs (s,i) and (i,t) allowing i to be the first or the last task of a sequence, respectively.

This graph is acyclic as illustrated in Figure 1.

In terms of this precedence graph, the problem is to determine a partition of the nodes into a set of paths (or sequences), each of these starting at s and ending at t, to minimize an objective function.

The problem can be formulated as a minimum cost flow problem defined on the precedence graph as follows :

<u>Problem (P)</u>

$$\text{Min} \sum_{(i,j)\varepsilon A} c_{ij}\, x_{ij}$$

$$\text{Subject to} \quad \sum_{i\varepsilon N} x_{ij} = 1 \qquad j\varepsilon N-\{s,t\}$$

$$\sum_{j\varepsilon N} x_{ij} = 1 \qquad i\varepsilon N-\{s,t\}$$

$$x_{ij} = 0 \text{ or } 1 \qquad (i,j)\varepsilon A$$

where $c_{ij}$ is the cost of traversing arc (i,j) and the decision variables

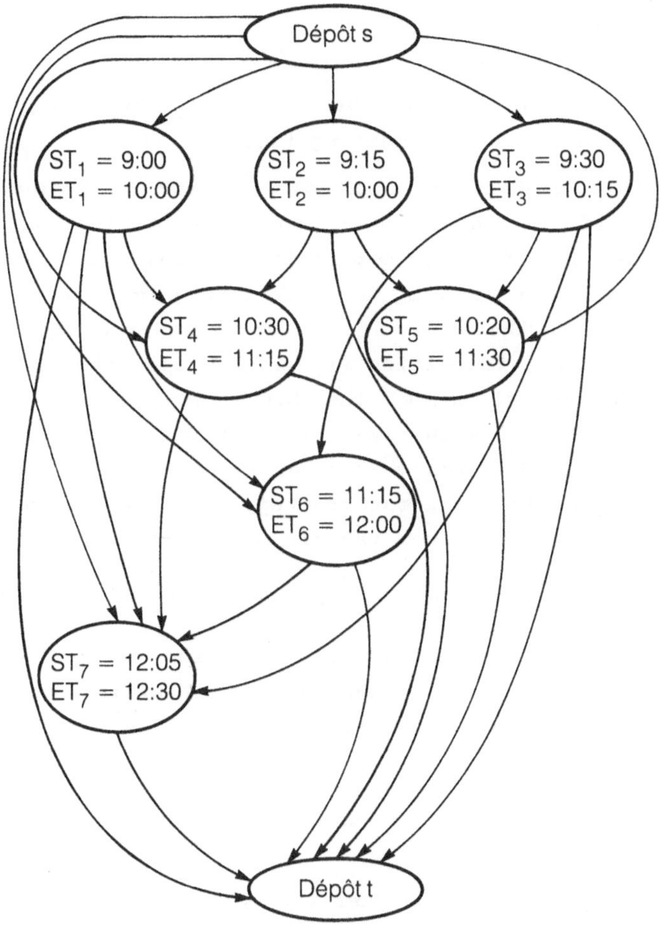


FIGURE 1

$$x_{ij} = \begin{cases} 1 & \text{if some vehicle traverses arc } (i,j) \\ 0 & \text{otherwise.} \end{cases}$$

The linear costs $c_{ij}$ are specified according to the objective of the problem. For instance, if the objective is to minimize fleet size, then

$$c_{ij} = \begin{cases} 1 & \text{if } i = s \\ 0 & \text{if } i \neq s, \end{cases}$$

and if the objective is to minimize fleet size and deadheading time, then

$$c_{ij} = \begin{cases} t_{ij} & \text{if } i \text{ and } j \text{ are different from } s,t \\ K & \text{if } i = s \\ 0 & \text{if } j = s \end{cases}$$

where the magnitude of K relative to the $t_{ij}$ measures the importance of fleet size versus deadheading time.

In general, this problem can be solved using a minimum cost flow algorithm. But if the objective is to minimize fleet size, then it reduces to Dilworth's chain decomposition problem and it can be solved with a variant of the maximal flow labeling procedure (see Ford and Fulkerson, 1962). Furthermore, if the start and end locations of all tasks are located in a few cities (i.e., the same city is the start or end location of several tasks) the following procedure due to Saha (1970) can be used : for each city, order chronologically the arrivals and departures from it, and assign arriving vehicles to departures according to a first-in-first-out strategy; when no vehicle is available for a departure, take a new vehicle out of the depot.

## 3. TIME WINDOWS

In Section 2 it was assumed that $ST_i$ is fixed for all tasks i. Now suppose that for each task i an earliest starting time $a_i$ and a latest starting time $b_i$ are specified :

$$a_i \leq ST_i \leq b_i.$$

The time interval $[a_i, b_i]$ is called the <u>time window</u> associated with task i.

A mathematical formulation is derived for this problem using the following oriented graph (N,A) :

i) With each task i associate a node i in N. With the depot, associate two nodes s and t.

ii) An arc (i,j) exists (i.e., $(i,j) \varepsilon A$ or task j might follow task i) if

$$a_i + D_i + t_{ij} \leq b_j.$$

Furthermore, for each task i, arcs (s,i) and (i,t) exist.

In terms of this oriented graph, the problem is to partition the nodes into a set of paths (or sequences), each of these starting at s and ending at t, to minimize an objective function such that the time window constraints are satisfied for each task. It is summarized as follows :

<u>Problem (PW)</u> Min $\sum_{(i,j)\varepsilon A} c_{ij} x_{ij}$

Subject to

$$\sum_{i\varepsilon N} x_{ij} = 1 \qquad j\varepsilon N-\{s,t\}$$

$$\sum_{j\varepsilon N} x_{ij} = 1 \qquad i\varepsilon N-\{s,t\}$$

$$x_{ij} = 0 \text{ or } 1 \qquad (i,j)\varepsilon A$$

$$x_{ij} > 0 \Rightarrow ET_i + t_{ij} \leq ST_j \qquad (i,j)\varepsilon \bar{A} \qquad (3.1)$$

$$a_i \leq ST_i \leq b_i \qquad i\varepsilon N-\{s,t\} \qquad (3.2)$$

where $\bar{A} = \{(i,j)\varepsilon A : i \neq s,t, \quad j \neq s,t\}$, $c_{ij}$ is the cost of traversing arc (i,j) and the decision variables

$$x_{ij} = \begin{cases} 1 & \text{if some vehicle traverses arc } (i,j) \\ 0 & \text{otherwise.} \end{cases}$$

Constraints (3.1) and (3.2) are referred to as the time window constraints.

Several methods have been proposed to solve this problem. Some of these are heuristic methods generating "good" solutions and are reviewed in Section 4. Other methods generate optimal solutions using "branch-and-bound" procedures. Some of these are reviewed in Section 5.

## 4. HEURISTIC METHODS

Four different heuristic methods are now summarized. The first method, proposed by Levin (1971) and by Swersey and Ballard (1984), solves a discrete approximation of problem (PW) specified in Section 3. The second method due to Orloff (1976) is a two phase procedure : phase 1 generates a feasible solution and phase 2 improves on it. The third method, proposed by Bodin and Berman (1979) and by Desrosiers et al. (1980), finds a feasible solution by solving a sequence of transportation problems. The fourth heuristic, presented in Graham and Nuttle (1984), generates a feasible solution using problem (PW) where the time window constraints are relaxed (or deleted).

### 4.1 Discrete approximation

The time window interval of each task is discretized in the sense that specific moments in the time interval are selected as candidate start times for the task. To derive the model, replace each variable $ST_i$ in Problem (PW) by a set of binary decision variables $y_\nu$ associated with candidate start times for task i. With these variables we specify a mathematical model very similar to Problem (P) in Section 2. Indeed consider the following precedence graph $(\tilde{N},\tilde{A})$ :

i) With each of the specific candidate starting times of each task, associate a node $\nu$. s and t are nodes associated with the depot.

ii) An arc $(\tau,\nu)$ exists if a vehicle can execute task j starting at the specific time $\nu$ after task i starting at specific time $\tau$. Also, for each $\nu\varepsilon\tilde{N}-\{s,t\}$, the arcs $(s,\nu)$ and $(\nu,t)$ exist.

Denote by $\Gamma_i = \{\nu\varepsilon N : \nu$ is candidate start time of task $i\}$.

The mathematical model is summarized as follows :

$$\text{Min} \quad \sum_{(\tau,\nu)} c_{\tau\nu}\, x_{\tau\nu}$$

$$\begin{array}{lll}
\text{Subject to} & \sum_{\tau\varepsilon\tilde{N}} x_{\tau\nu} = y_\nu & \nu\varepsilon\tilde{N}-\{s,t\} \\
& \sum_{\tau\varepsilon\tilde{N}} x_{\nu\tau} = y_\nu & \nu\varepsilon\tilde{N}-\{s,t\} \\
& x_{\tau\nu} = 0 \text{ or } 1 & (\tau,\nu)\varepsilon\tilde{A} \\
& \sum_{\nu\varepsilon\Gamma_i} y_\nu = 1 & 1\leqslant i\leqslant m \\
& y_\nu = 0 \text{ or } 1 & \nu\varepsilon\tilde{N}-\{s,t\}
\end{array}$$

where $c_{\tau\nu}$ is the cost of traversing arc $(\tau,\nu)$ and the decision variables

$$x_{\tau\nu} = \begin{cases} 1 & \text{if a vehicle traverses arc } (\tau,\nu) \\ 0 & \text{otherwise.} \end{cases}$$

Swersey and Ballard (1984) have used this model in the context of school busing where the objective is to reduce the fleet size. In this paper, they report that the optimal solution for the linear programming relaxation of the model where constraints $x_{\nu\tau} = 0$ or $1$ are replaced by $0\leqslant x_{\nu\tau}\leqslant 1$ and constraints $y_\nu = 0$ or $1$ are replaced by $0\leqslant y_\nu\leqslant 1$, was integer in many of their examples.

## 4.2 Matching methods

To specify the two phase method due to Orloff (1976), consider the following oriented graph :

i) With each task i associate a node i.

ii) With each pair of nodes i and j, associate two arcs :

. arc (i,j) with cost

$$c_{ij} = \begin{cases} t_{ij} & \text{if } a_i + D_i + t_{ij} \leq b_j \\ K & \text{otherwise} \end{cases}$$

. arc (j,i) with cost

$$c_{ij} = \begin{cases} t_{ji} & \text{if } a_j + D_j + t_{ji} \leq b_i \\ K & \text{otherwise.} \end{cases}$$

During the first iteration of phase 1, a minimum weighted matching problem is solved on the preceding graph. This is essentially equivalent to identifying a set of arcs such that each node is the extremity of one and only one arc (i.e., no two arcs have a node in common) and the sum of their cost is minimized. If an arc (i,j) belongs to this set, and if $c_{ij} < K$, then collapse nodes i and j into one node i' and specify accordingly a time window interval $[a_{i'}, b_{i'}]$ for this pair of tasks assumed to be collapsed into one. Then specify a new oriented graph with some of the tasks collapsed into one, and repeat the procedure.

When no further matchings are possible (i.e., when the optimal solution of the minimum weighted matching problem includes only arcs with cost K), each node of the current graph corresponds to the schedule of a vehicle executing the sequence of tasks collapsed at this node. This solution is feasible for problem (PW), but is not optimal in general. Assume that z vehicles are necessary for this solution.

During phase 2, the solution generated in phase 1 is improved. Consider the oriented graph (N,A) specified in Section 3, but modify it to include z nodes associated with the depot, each with arcs to and from every task node. Also, add arcs to and from every depot node and every other depot node with cost 0. The solution generated in phase 1 corresponds to a travelling salesman tour in this augmented graph if we assume that each depot node is the starting node of one and only one vehicle and the ending node of one and only one other vehicle. The exchange procedure proposed in Lin (1965) can be adapted to improve this initial tour.

This procedure always yields a feasible solution which is not, in general, optimal.

### 4.3 Assignment method

To specify the method, we refer to the oriented graph (N,A) introduced in Section 3, but the depot nodes will not be relevant. Denote $N' = N-\{s,t\}$.

At each iteration of the procedure, a transportation problem is specified in terms of two subsets of tasks, $N_1$ and $N_2$. The subset $N_1$ is the set of source-tasks including the tasks already executed by a vehicle which is now available for other tasks. The subset $N_2$ includes a subset of sink-tasks to be executed (i.e., for which a vehicle is required).

For the first iteration, the subset $N_1$ includes the tasks in N' that cannot be executed after any other task. Hence $j \epsilon N_1$ if $a_i + D_i + t_{ij} > b_j$ for all $i \epsilon N'$, $i \neq j$. Also, if two tasks can follow each other but cannot be executed after any other task, then select one of these and include it in $N_1$. Hence each task in $N_1$ should be the first task of a different sequence. Furthermore, once a task in $N_1$ is executed, the vehicle is available for other tasks. The sequences are built up by solving a sequence of transportation problems. To determine the subset $N_2$ of sink-tasks to be executed next, replace N' by $N'-N_1$, and repeat the preceding procedure used to define $N_1$.

For each pair of nodes $i \epsilon N_1$, $j \epsilon N_2$, define an asignment cost $ac_{ij}$ for using a vehicle to execute j after i as follows :

$$ac_{ij} = \begin{cases} t_{ij} & \text{if } a_i + D_i + t_{ij} \leq b_j \\ K_1 & \text{otherwise} \end{cases}$$

where $K_1 \gg K$. Introduce a dummy source $\bar{s}$ and a dummy sink $\bar{t}$. For all $i \epsilon N_1$ define an assignment cost $ac_{i\bar{t}} = 0$ to indicate that the vehicle becomes available after executing task i. For all $j \epsilon N_2$ define an assignment cost $ac_{\bar{s}j} = K$ to indicate that a new sequence is initiated with task j.

This transportation problem is then solved. If $i \epsilon N_1$ is assigned to $\bar{t}$, then i remains in set $N_1$. If $\bar{s}$ is assigned to $j \epsilon N_2$, then a new sequence is initiated and the vehicle is available after executing task j, and j is introduced into $N_1$. If $i \epsilon N_1$ is assigned to $j \epsilon N_2$, then the vehicle executes j after executing i (i.e. i and j belong to the same sequence). Then i is eliminated from $N_1$, j is introduced into $N_1$ because the vehicle is no longer available after task i but rather after task j. Furthermore the earliest time $a_j + D_j$ at which the vehicle is available is adjusted accordingly to the sequence to which it belongs. Also we make the assumption that each task of a sequence is executed as early as possible.

The procedure is repeated. The current set $N_1$ of source-tasks is used. The set $N_2$ is determined using the same procedure with N' replaced by the set of node-tasks that have not been assigned to any sequence as yet. The process stops when this set is empty.

The method yields a feasible solution which is not optimal in general.

### 4.4 Relaxation method

To use this method, Graham and Nuttle (1984) make the assumption that the oriented graph (N,A) is acyclic. This assumption is acceptable in most applications where task durations are longer than the length of the time window interval.

First, problem (PW) where the time window constraints (3.1) and (3.2) are relaxed (deleted) is solved. Note that the relaxed problem reduces to an assignment problem. The optimal solution generates a partition of the nodes into a set of paths (sequences). Under the assumption that the graph is acyclic, each path starts at s and ends at t. But some of these sequences might be infeasible in the sense that the time window constraint is violated for at least one task of the sequence. If all sequences are feasible, the solution is optimal. Otherwise, the sequences have to be modified to recover feasibility. In their implementation of the method, the adjustments of the sequences are carried out manually by experienced scheduling personnel.

## 5. EXACT METHODS

In this secion we summarize branch-and-bound exact methods to find optimal solutions to problem (PW). The first method presented by Desrosiers et al. (1984) uses a column generation technique to solve a set partitioning problem equivalent to problem (PW). The second method requires the resolution of a problem of type (PW) where the time window constraints are relaxed at each node of the branch-and-bound tree until an optimal solution is reached. It is presented in Desrosiers et al. (1983a).

### 5.1 Column generation technique

Assume that $\Omega$ is a set of sequences of tasks, and that each of these sequences satisfies the time window constraints. Use the decision variables $y_r$, $r \in \Omega$ to specify the following set partitioning problem:

Problem (SP) Min $\sum_{r\varepsilon\Omega} c_r\, y_r$

Subject to $\sum_{r\varepsilon\Omega} \delta_{ri}\, y_r = 1 \qquad 1 \leq i \leq n$

$y_r = 0 \text{ or } 1 \qquad r\varepsilon\Omega$

where

$$y_r = \begin{cases} 1 & \text{if sequence } r\varepsilon\Omega \text{ is used} \\ 0 & \text{otherwise.} \end{cases}$$

$$\delta_{ri} = \begin{cases} 1 & \text{if task } i \text{ belongs to sequence } r \\ 0 & \text{otherwise,} \end{cases}$$

and $c_r$ is defined according to the $c_{ij}$ and is the cost for a vehicle to execute the $r^{th}$ sequence of tasks.

At each iteration of the procedure the linear programming relaxation of a problem (SP) associated with a given $\Omega$ is solved. This relaxation is obtained by replacing the constraints $y_r = 0$ or 1 by $y_r \geq 0$. Denote by $\sigma_i$ the optimal multiplier associated with constraint $\sum_{r\varepsilon\Omega} \delta_{ri}\, y_r = 1$. Pricing out the $r^{th}$ sequence, we obtain the relative cost

$$\sum_{(i,j)\varepsilon A} \delta_{ri}\, \delta_{rj}\, c_{ij} - \sum_{i\varepsilon N} \delta_{ri}\, \sigma_i.$$

Hence the approach is to generate new sequences not already in $\Omega$ which might reduce the overall cost of the set partitioning problem. The new sequences have negative relative costs when evaluated with the optimal multipliers $\sigma_i$. It is easy to see that such a sequence is the optimal solution to a shortest path problem with the additional time window constraints in the oriented graph (N,A) when the cost of arc (i,j) is taken equal to $(c_{ij} - \sigma_i)$. Desrosiers et al. (1983b) propose an algorithm to solve this problem.

This procedure is embedded into a branch-and-bound method to obtain an optimal integer solution to a problem (SP) equivalent to (PW). At a node of the branch-and-bound tree, if the linear programming relaxation of the set partitioning problem has a fractional optimal solution, then branching takes place on the sequence for which the associated decision variable can be set to 1 at the least additional cost according to the estimates given by the marginal costs. Furthermore, the branching is completed according to the Bellmore and Malone (1971) method as adapted by Carpaneto and Toth (1980) for the travelling salesman problem. This is illustrated in Figure 2 where we assume that sequence r includes tasks $i_1, i_2, \ldots, i_m$. The first branch imposes the existence of sequence r. The second branch imposes the sequence $i_1$, $i_2$, ..., $i_{m-1}$ but prohibits that $i_m$ follows $i_{m-1}$, and so on. The next to last branch imposes the sequence $i_1$, $i_2$ but prohibits that $i_3$ follows $i_2$. The last branch prohibits that $i_2$ follows $i_1$.

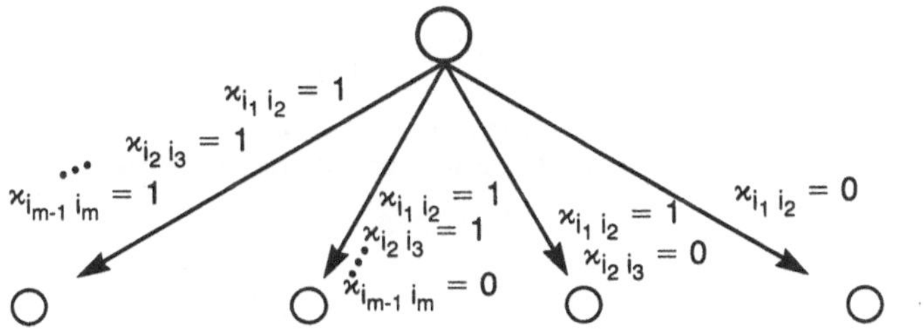


FIGURE 2

## 5.2 Time-window constraints relaxation

As in Section 4.4, consider problem (PW) where the time window constraints are relaxed (i.e., discarded). Sequences of tasks can be generated by solving this relaxed version of (PW). If the assumption that the graph (N,A) is acyclic is not verified, this optimal solution might include subtours where the vehicle can return to the first task of the sequence after executing the last task of the sequence. Furthermore, the time window constraints might not be satisfied for some tasks in the sequences generated. Of course, if neither of these cases occurs, then the solution is optimal for (PW).

On the other hand, to eliminate subtours and sequences where some time window constraints are not satisfied, a branch-and-bound procedure is used where a relaxed version of a problem of type (PW) is solved at each node of the branch-and-bound tree.

Two branching procedures are presented by Desrosiers et al. (1983a). They are applied only on the infeasible part of the sequence including the tasks for which the time window constraints of the last task of the subsequence cannot be satisfied even in the first task of the subsequence is executed as early as possible.

This notion of infeasible subsequence is illustrated in Figure 3 where the subsequence of tasks 3, 4, 5 and 6 is not feasible because the time-window constraint of task 6 is violated even if 3 is initiated at $a_3$. Indeed, to be able to initiate task 6 at $b_6$ (latest starting time for 6), task 3 should be initiated no later than $LT_3$ which is earlier than $a_3$.

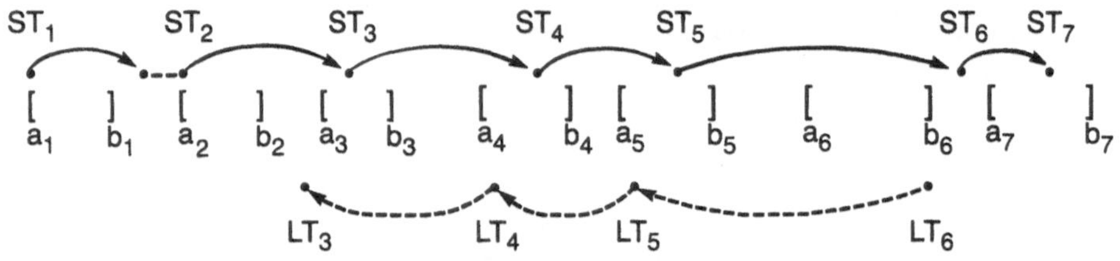


$ST_i$ is the starting time of task i in the sequence.
$LT_i$ is the latest starting time of task i according to the sequence.

FIGURE 3

As in Section 5.1, the branching procedures proposed by Desrosiers et al. (1983a) represent an adaptation of the Bellmore and Malone (1971) approach. The first procedure of branching on the flow variables is very similar to the procedure used in Section 5.1. It is illustrated in Figure 4 to eliminate the infeasible subsequence of tasks 3, 4, 5 and 6 of the preceding example. The first branch imposes subsequence 3, 4, 5 and prohibits the link 5,6. The second one imposes subsequence 3, 4 and prohibits link 4,5. Finally, the last branch prohibits link 3,4.

The second branching procedure is specified in term of the time subintervals of the tasks of the infeasible subsequence. These are specified according to the notation introduced in Figure 3. For each task i of the infeasible subsequence, denote

- $L_i$ the subinterval $[a_i, LT_i]$
- $R_i$ the subinterval $(LT_i, b_i]$.

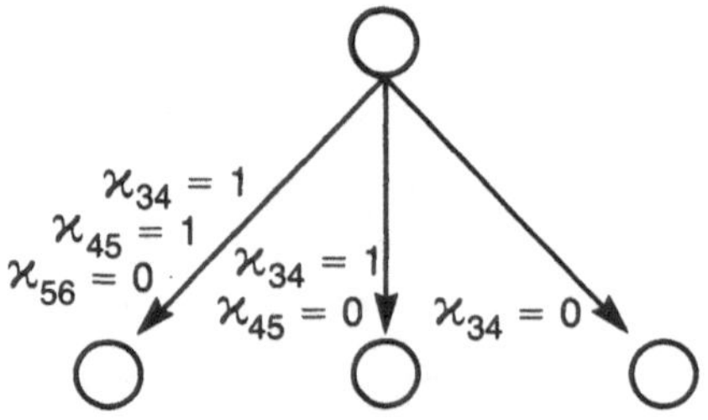


FIGURE 4

The branching procedure illustrated in Figure 5 generates branches where some tasks are required to be initiated within one or the other subinterval. For the preceding example it is easy to verify that links (3,4), (4,5) and (5,6) are prohibited by the first, the second and the third branch, respectively.

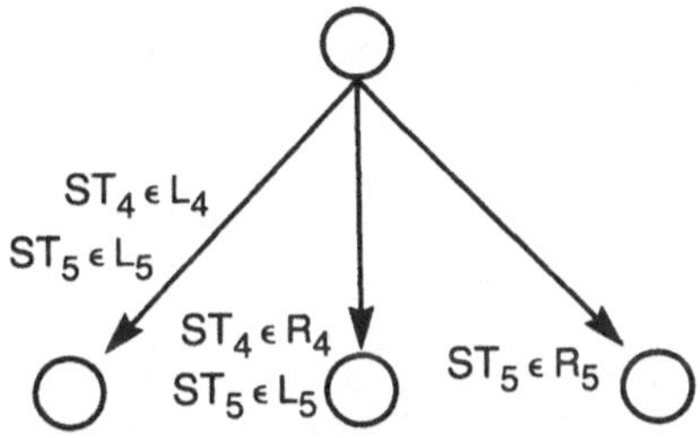


FIGURE 5

Minor adjustments are required to use these procedures to eliminate subtours where time-window constraints are satisfied for all tasks. For the details we refer the reader to Desrosiers et al. (1983a).

The branching strategy selecting the subtour or illegal subsequence on which branching takes place is specified in terms of a bound on the increase in the optimal value for the relaxed problem when a link is prohibited. The subtour, or illegal subsequence, generating the least number of branches where this bound is positive, is selected.

## 6. CONCLUSION

The purpose of this paper was to specify the one depot vehicle scheduling problem and to introduce several approaches to analyzing the problem with time window constraints. The paper does not report any numerical comparison of the relative efficiency of the methods because the author is not aware of any study of this sort, and because the use of any specific method depends on the context in which the problem arises. Of course, the exact methods are more time consuming than heuristics, but if the objective is to reduce the fleet size to a minimum, the savings of the costs of a vehicle might justify the use of exact methods.

The presentation of the different methods is very sketchy to underline only the basic principles. The reader is referred to the specific papers to see how the authors implement these basic principles and use other techniques and principles to accelerate these methods.

REFERENCES

BELLMORE, M. and MALONE, J.C. (1971). "Pathology of Travelling Salesman Subtour Elimination Algorithm", Operations Research 19, 278-307.

BODIN, L. and BERMAN, L. (1979). "Routing and Scheduling of School Buses by Computer", Transportation Research 13, 113-129.

BODIN, L., ASSAD, A., BALL, M. and GOLDEN, B. (1983). "The State of the Art in Routing and Scheduling of Vehicles and Crews", Computers and Operations Research 10, 69-211.

CARPANETO, G. and TOTH, P. (1980). "Some New Branching and Bounding Criteria for the Asymmetric Travelling Salesman Problem", Management Science 26, 736-743.

DESROSIERS, J., FERLAND, J.A., ROUSSEAU, J.-M., LAPALME, G., CHAPLEAU, L. (1980). "TRANSCOL : A Multi-Period School Bus Routing and Scheduling System", Publication #164, Centre de recherche sur les transports, Université de Montréal, Montréal, Canada; also in Ignall, E.J. and Swersey, A.J. (editors), Management Science and the Delivery of Urban Services, Elsevier-North-Holland (forthcoming).

DESROSIERS, J., SOUMIS, F. and DESROCHERS, M. (1984). "Routing with Time Windows by Column Generation", Networks 14, 545-565.

DESROSIERS, J., SOUMIS, F. and DESROCHERS, M. (1983a). "Routing and Scheduling by Network Relaxation and Branch-and-Bound on Time Variables", Publication #278, Centre de recherche sur les transports, Université de Montréal, Montréal, Canada; also in Computer Scheduling of Public Transport, Volume II, North-Holland (forthcoming).

DESROSIERS, J., PELLETIER, P. and SOUMIS, F. (1983b). "Plus court chemin avec contraintes d'horaire", RAIRO 17, 357-377.

FORD, L.R. and FULKERSON, D.R. (1962). "Flows in Networks", Princeton University Press, Princeton, New Jersey.

GRAHAM, D. and NUTTLE, L.W. (1984). "A Comparison of Heuristics for a School Bus Scheduling Problem", IE Technical Report No. 84-3, Department of Industrial Engineering, North Carolina State University, Raleigh, North Carolina.

LEVIN, A. (1971). "Scheduling and Fleet Routing Models for Transportation Systems", Transportation Science 5, 232-255.

LIN, S. (1965). "Computer Solutions of the Travelling Salesman Problem", Bell System Technical Journal 44, 2245-2269.

ORLOFF, C. (1976). "Route Constrained Fleet Scheduling", Transportation Science 10, 149-168.

SAHA, J. (1970). "An Algorithm for Bus Scheduling Problems", Operational Research Quarterly 21, 463-474.

SWERSEY, A.J. and BALLARD, W. (1984). "Scheduling School Buses", Management Science 30, 844-853.

# PROBABILISTIC ANALYSIS OF ALGORITHMS FOR SOME COMBINATORIAL OPTIMIZATION PROBLEMS

P.G. Gazmuri
Depto. Ing. de Sistemas
Pontificia Universidad Católica de Chile
Casilla 6177, Santiago, Chile

## 1. INTRODUCTION

In this paper we consider some NP - complete combinatorial optimization problems in the field of machine scheduling and graph coloring. In short, a problem being NP - complete means that it is very unlikely that a polynomial time optimizing algorithm will ever be found for its solution. A probabilistic analysis of algorithms for these problems is presented. Under this approach, it is assumed that the instances of the problem are drawn from some reasonable probability distribution, although the data specifying an instance is known before the algorithm is applied. The behaviour of an algorithm can then be analyzed as some sort of stochastic process. Usually, the asymptotic performance (as the problem size increases) is emphasized.

The probabilistic analysis of algorithms has attracted the attention of many researchers during the last years. It was initiated with the pioneer work of R.M. Karp [9] on the travelling salesman problem. More details on this approach can be found in [2], [10].

In the next two sections of the paper a collection of some of the author's results are sketched. Details on the proofs of such results are omitted and can be found on the references cited.

The machine scheduling problem to be studied is the following: there are n jobs $J_1, J_2, \ldots, J_n$ that have to be processed on one available machine. Job $J_i$ arrives at epoch $R_i$ and has $P_i$ units of processing time. The machine can handle at most one job at a time and no preemptions are allowed. We want to minimize the sum of the completion times of all jobs. This problem is known as $1/ R_j / \Sigma C_j$ in the scheduling literature. Our first result is concerned with the SPT (Shortest Processing Time) rule: we show that instances of the problem can be constructed for which the error of the solution obtained by such rule (relative to the value of an optimal solution) is as big as desired. As a counterpart, we then show that for certain type of probability distributions of the instances, the SPT rule is asymptotically optimal, in the sense that the error of the solution tends to 0 as n goes to ∞. Finally, we consider the generalization of the problem to the case of m parallel identical machines. Two asymptotically optimal algorithms are presented which are extensions of algorithms previously obtained for the case of one machine.

There are some interesting published results on probabilistic analysis of scheduling algorithms. Loulou [11] uses this approach for the problem of minimizing the completion time of n jobs on m parallel machines. Also, there are some results on stochastic scheduling that are somewhat related to the approach used in this paper. Pinedo [12] and Weiss [13] consider a family of scheduling problems under a different probabilistic framework: they assume that the processing times of jobs are known only in distribution (and not the actual values) and then optimal policies are considered that optimize some objective function in expectation.

The graph optimization problem considered in the last part of the paper is the chromatic number problem which is defined as follows: given an undirected graph G = (N, E), where N is the node set and E the arc set, we want to determine the minimum number of colors needed to color the nodes in G so that adyacent nodes are assigned different colors. The probabilistic analysis presented focuses on the case of random sparse graphs, in which the average number of edges is a linear function of the number of nodes. The random dense model has been considered previously by Grimett and Mc Diarmid [8]. We analyze the asymptotic behaviour of a simple sequential coloring algorithm and present upper and lower bounds for the chromatic number of the graph.

## 2. MACHINE SCHEDULING

We want to consider now the $1/ R_j / \Sigma C_j$ scheduling problems defined in section 1. If I is an instance of it, let $v^*(I)$ be the value of an optimal solution and $v_{SPT}(I)$ the value of the solution obtained using the SPT rule. This rule operates as follows: every time a job is completed, we choose as the next job to be processed the one with shortest processing time in the queue. The next proposition states that there are instances for which SPT is as bad as desired.

PROPOSITION 1: Let $e > 0$ be any real number. Then there exists an instance $I(e)$ such that

$$\frac{v_{SPT}\ I(e) - v^*\ I(e)}{v^*\ I(e)} > e$$

The details on how $I(e)$ is contructed appear in [5]. Note that this result has no probabilities involved.

We now review some previous results on the probabilistic analysis of the problem, that will be needed to present our arguments. Firstly, the probability distribution of the instances of the problem is specified as follows:

- the processing times $P_1, P_2, \ldots, P_n$ are assumed to be i.i.d. random variables, bounded above by some constant M.
- we define $T_1 = R_1$, $T_i = R_i - R_{i-1}$, $i = 2\ldots n$ as the inter-arrival times and we assume that they are i.i.d. random variables.
- the sequences $\{P_i\}$ and $\{T_i\}$ are independent.
- the saturation coefficient $\rho$ is defined as $\rho = E(P_i)/E(T_i)$. The instances of the problem can be divided into two categories: the undersaturated case ($\rho < 1$) and the oversaturated case ($\rho > 1$). An asymptotically optimal algorithm has been previously developed for each of them. The basic structure of these algorithms is briefly presented next.

Undersaturated Case ($\rho < 1$)

The key idea in this case is that of a conservative solution. In such a solution, the machine is never idle when there are jobs waiting to be processed. If we examine the structure of such a solution in a time axis it will look as follows:

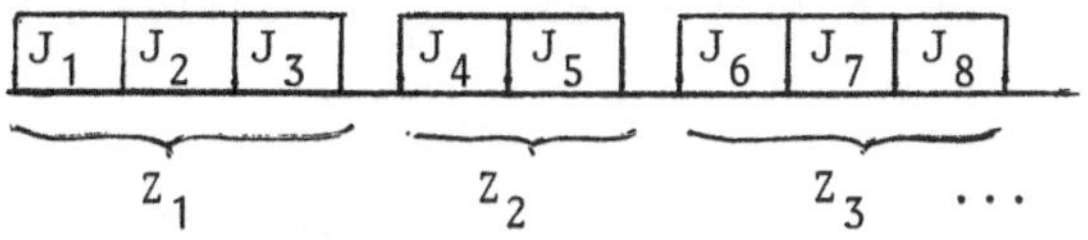

Groups of jobs will be processed consecutively and natural idle times will occur between them. If we call $Z_1$, $Z_2$, $Z_3$ .. the number of jobs in the groups, it can be shown [4] that $E(Z_i)$ is a constant, independent of n, the number of jobs. Unfortunately; it cannot be guaranteed that an optimal solution is conservative (i.e. there are instances of the problem for which, even though there are jobs waiting to be processed, it is convenient to keep the machine idle until the next job arrives). This means that an optimal solution is likely to contain natural and artificial idle times, the last ones being produced by the solution procedure itself.

In [4] an asymptotically optimal algorithm called INSERTION is presented for this case. Broadly speaking, it works as follows:

i) Compute an optimal solution for the first $a = \lfloor \ln \cdot n \rfloor$ jobs (this can be done in at most $O(n)$ time). The form of such a solution will be the following:

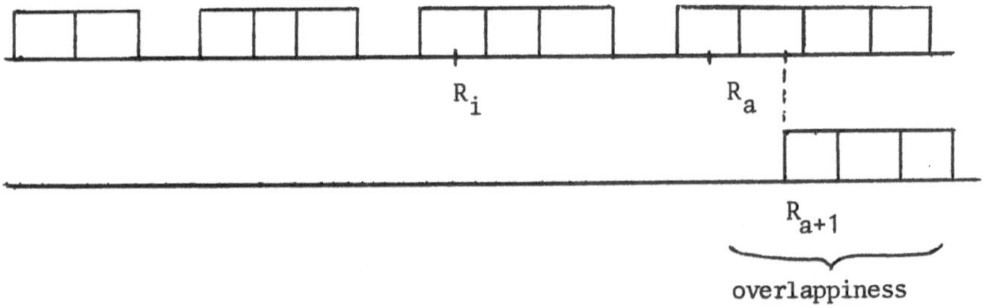


The first time axis contains the solution. Because of the artificial idle times, there will be some overlappiness of this optimal solution with jobs $J_{a+1}$, $J_{a+2}$, etc. (represented in the second time axis).

ii) An arrival point $R_i$, $i < a$ is selected towards the end of the optimal schedule for the first a jobs. Jobs in the queue, waiting to be processed at that time, are left aside. A conservative solution is gnerated from then on (involving jobs $J_{i+1}$, $J_{i+2}$, etc.) until the sum of the natural idle times so generated is at least as equal as the sum of the processing times of the jobs left aside. $R_i$ is called a breaking point.

iii) The jobs $J_{i+1}$, $J_{i+2}$, etc. in the conservative solution are pushed to the right so that no idle time is left between them. This will produce one big idle time from $R_i$ on, where the jobs left aside can be processed (conservatively).

iv) Repeat steps i), ii) and iii) for the next a jobs, and so on.

The following result can be proven concerning the error of the algorith [4].

Theorem 2. If for a problem with n jobs we let v*(n) and v(n) be the value of an optimal solution and the value of the solution provided by INSERTION respectively, then with probability $1 - 0\ [(\ln n)^2/\ n]$ it holds that

$$\frac{v(n) - v^*(n)}{v^*(n)} = 0\ [(\ln \ln n)^{-\alpha}], \quad 0 < \alpha < 1$$

From this it follows that, with probability going to 1 as $n \to \infty$, the error of the algorithm goes to 0 (although the convergence rate is very slow). The proof of the theorem relies on two main facts:

- it can be shown that, with high probability, v*(n) is 0(n) (0(ln n) for a jobs).
- the breaking point $R_i$ can be conveniently chosen so that the error induced by the algorithm for the first a job is o(ln n). This considers the effect of the conservative (and hence not necessarily optimal) solution for jobs $J_{i+1}$, $J_{i+2}$,..., etc. and the increase in their waiting times due to being pushed to the right.

Oversaturated Case ($\rho > 1$)

The algorith developed for this case differs strongly from INSERTION. It involves two major steps:

- first, an optimal solution is obtained for the case when preemptions of jobs are allowed. It is well known that the residual SPT rule is optimal in such a case [7]. Under this rule, if the processing time of an arriving job is less than what is left to be processed of the job in the machine, then a preemption occurs. Upon completion of a job, the one in the queue with shortest processing time is selected for processing.
- a feasible solution is then obtained by means of a collection of patching operations, by which all portions of a job are put together.

Suppose job $J_i$ was preempted 2 times in the optimal preemptive solution. The effect of the patching operations on $J_i$ is depicted graphically in the following figure:

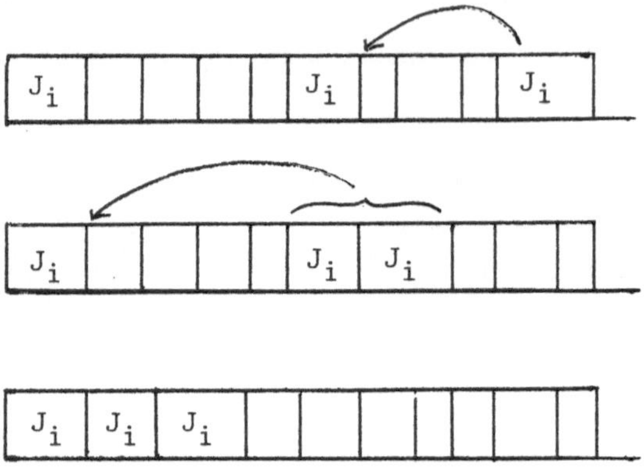


This algorithm is called PREEM-PATCH. The following result is proven in [4].

Theorem 3. For a problem with n jobs, let $v^*(n)$ be the value of an optimal solution and $v(n)$ be the value of the solution provided by PREEM-PATCH. Then, with probability $1-0(n^{-1})$ it holds that:

$$\frac{v(n) - v^*(n)}{v^*(n)} = 0(n^{-1})$$

Two facts are used for proving this result:

- with high probability, $v^*(n)$ is $0(n^2)$
- the expected increase in waiting time of jobs that get processed in between portions of $J_i$ is constant, so the overall effect is at most $0(n)$.

Note that the convergence rate of the error in Theorem 3 is much faster than for the INSERTION algorithm. Also, it is easy to realize that the solution provided by PREEM-PATCH differs from the SPT solution. The following theorem, which is a new result, is a definite statement on how both solutions compare. A complete proof can be found in [5].

Theorem 4. Let I be any instance of the $1/ R_j / \Sigma C_j$ problem and let $v(I)$ and $v_{SPT}(I)$ represent the value of the PREEM-PATCH and SPT solution for I, respectively. Then:

$$v(I) \geq v_{SPT}(I).$$

A consequence of Theorems 3 and 4 is that the SPT rule is asymptotically optimal for the oversaturated case.

To prove Theorem 4, one starts with the solution obtained using the residual SPT rule. Then the patching operations are performed (as specified in the PREEM-PATCH algorithm) but, for some specific jobs, a reordering operation using the SPT rule is applied after each patching operation. It can be shown that if this is done, 2 things happen:

- the reordering operations do not introduce any increase in the value of the solution.
- at the end of this procedure, the (feasible) solution obtained will be precisely the one it would have been produced if the SPT rule was applied. From this, the result follows. Note that the statement in Theorem 4 is purely combinatorial.

The $m/R_j/\Sigma C_j$ scheduling problem

We now consider the extension of the $1 /R_j / \Sigma C_j$ to the case of m parallel identical machines. $P_i$ represents the procesing time of job $J_i$ in anyone of the machines. The probability distribution of the instances is specified in the same way as before. The saturation coeficient is now defined as $\rho = E(P_i)/mE(T_i)$. Generalizations of INSERTION and PREEM-PATCH are presented next.

Undersaturated Case ($\rho = E(P_i)/mE(T_i) < 1$)

The INSERTION algorithm can be extended to the case of m parallel machines as follows:

i) compute an optimal solution for $m/R_j/\Sigma C_j$ when only the first a jobs are considered ($a = [\ln n]$).

ii) choose an arrival point $R_i$, $i < a$ as a breaking point. Let $A(R_i)$ be the set of jobs present in the system at that time, including the (possibly) m jobs being processed.

iii) a random reordering of jobs in $A(R_i)$ is obtained. These jobs are then assigned cyclically to machines 1 through m (first job to machine 1, second job to machine 2,..., job (m+1) to machine 1, etc.). Let $A_k(R_i)$ be the set of jobs assigned to machine k and $Q_k(R_i) = |A_k(R_i)|$. Let $S_k(R_i)$ be defined as:

$$S_k(R_i) = \sum_{J_j \in A_k(R_i)} P_j$$

iv) Jobs $J_{i+1}, J_{i+2},\ldots$, etc. are assigned cyclically to machines 1 through m and a conservative solution is generated at each machine. This is done until the sum of natural idle times at machine k is $> S_k(R_i)$, $k = 1, 2,\ldots, m$. Let $b_k(i)$ be the number of jobs needed to be processed conservatively to meet this goal. Let $b(i) = b_1(i) + \ldots + b_m(i)$.

v) at machine k, k = 1, 2,..., m the following procedure is applied:

- starting at $R_{i+k}$, obtain a conservative solution for the $b_k(i)$ jobs from the set $\{J_{i+1},\ldots, J_{i+b(i)}\}$ that were assigned to machine k. Push these jobs to the right so that no idle times occur between them.
- schedule jobs in $A_k(R_i)$ using SPT, starting at time $R_{i+k}$.

vi) consider the next group of a jobs and repeat the previous steps, etc.

The following result is proven in [5]:

<u>Theorem 5</u>. Let $F(i) = (1/m)\ (M^2Q^2(R_i) + MQ(R_i)\ (a-i))$ and let $0 < \alpha < 1$, $0 < \varepsilon < 1/2$. If for each set of a jobs the breaking point $R_i^*$ is chosen, where:

$F(i^*) = \mathrm{Min}\{F(i) : a - a^{1/2+\varepsilon} < i < a\}$ then, with probability $1-0[(\ln n^2)/n]$, the error of the solution produced by INSERTION - m is $0\ [(\ln \ln n)^{-\alpha}]$.

<u>Oversaturated Case ($\rho = E(P_i)/m\ E(T_i) > 1$)</u>

In the case of m parallel machines, an optimal preemptive solution can easily be obtained using an extension of the residual SPT rule: an arriving job produces a preemption if its processing time is less than the remaining processing time of at least one job being processed at that instant. The structure of such a solution is rather complicated though since pieces of the same job may be processed at different machines. Hence, there is no simple rule for obtaining a feasible solution. The following strategy can be formulated to overcome this difficulty (we call this procedure SPT-m):

i) jobs $J_1, J_2, \ldots, J_n$ are assigned cyclically to machines 1 through m.

ii) apply the SPT rule independently to each machine.

Theorem 6 [5]. With probability $1 - O(n^{-1})$, the error of the solution provided by SPT-m is $O(n^{-1})$.

In proving this result, one only needs to compare the solution obtained by the cyclic assigment of jobs to the machine and the independent application of the residual SPT rule to each of them, with the solution obtained by applying the extension of the residual SPT rule to the whole problem. It can be shown that the difference in the value of these solutions is $O(n)$, while the value of an optimal solution is $O(n^2)$.

## 3. GRAPH COLORING

We consider now the chromatic number problem in a random sparse graph $G = (N, E)$. In this case, an edge is present with probability $p = 2c/(n-1)$, independently of the behavior of other edges (here, n is the number of nodes and c is a constant). It follows that the expected number of edges in G is cn the average degree of a node is 2c.

### A simple sequential coloring algorithm

In a feasible solution to the chromatic number problem, all the nodes to which a same color is assigned have the property that there are no arcs between them. Such a collection of nodes is called an independent set. Our main tool for generating a coloring procedure is a simple sequential algorith for constructing a maximal independent set, which works as follows: examine the nodes sequentially and, at each stage, include the current node if it has no arcs to the previously included nodes.

We have analyzed the asymptotic behaviour of this algorithm in a companion paper [6]. If we let $w_n$ be a random sparse graph with n nodes and $\sigma(w_n)$ the size of the independent set constructed by the sequential algorithm when applied to $w_n$, we have shown that, with probability tending to 1 as $n \to \infty$ one has:

$$\left\lceil \frac{\ln(2c+1-\varepsilon}{2c} n \right\rceil \leq \sigma(w_n) \leq \left\lfloor \frac{\ln(2c+1+\varepsilon)}{2c} n \right\rfloor$$

where $\varepsilon$ is a constant > 0.
(here $\lceil x \rceil$ is the smallest integer $\geq x$, and $\lfloor x \rfloor$ is the integer part of x).

Our overall strategic for coloring $w_n$ is a follows: we start by applying the sequential algorithm to $w_n$; the resulting independent set will constitute a color class (that is, nodes to which the same color is assigned). Removal of these nodes will leave us with a residual graph (which happens to be random) to which the sequential algorithm is applied, to construct a new color class. This process is repeated a given number of times (we will specify precisely how many). The residual graph obtained after an application of the algorithm will be sparser than its predecessor. After a while, we will be able to recognize a residual graph of simple structure, for which a simple coloring procedure can be given, that uses at most 3 colors. The underlying property of this graph is that the average degree of a node must be < 1.

Our first task is then to determine how many applications of the sequential algorithm are needed to bring the average degree below 1.

Theorem 7. Let X* be the number of applications of the sequential algorithm needed to obtain a residual graph with average degree node < 1.

If $\varepsilon < \text{Min}\{2c, 1/2\}$ then with probability tending to 1 as $n \to \infty$, it holds that:

$$X^* = \text{Min}\left\{ i : \sum_{j=1}^{i} X_j \geq 2c - 2\varepsilon \right\} \qquad (1)$$

where $\{X_j\}$ satisfy:

$$X_1 = \ln (2c + 1 - \varepsilon)$$

$$X_i = \ln (e^{X_{i-1}} - X_{i-1}).$$

If the average degree of a node in the graph is $< 1$, then the following statement can be made concerning its structure.

Proposition 8 [3]. Let $w_n$ be a random graph where each edge is present with probability $p = 2c/(n-1)$, with $c < 1/2$ (i.e. average node degree $< 1$). Then, with probability going to 1 as $n \to \infty$, the components of $w_n$ are either trees or components containing exactly one cycle.

It is clear that components containing exactly one cycle can be colored with at most 3 colors. Trees can be colored with 2 colors. A simple breadth first procedure can be applied to produce such a coloring. The sequential graph coloring algorithm can then be summarized as follows:

i) compute $X^*$ according to (1)

ii) apply the sequential algorithm for constructing independent sets $X^*$ times.

iii) apply a breadth first procedure to color the residual graph with at most 3 colors.

## A lower bound for the chromatic number

Let $X(w_n)$ be the chromatic number of $w_n$. A lower bound for $X(w_n)$ can easily be obtained if an upper bound for the maximum independent set in $w_n$ is known. Such a bound has been obtained in [6].

Here we will use a different approach, which focuses directly on the probabilistic structure of the graph.

We will say that the collection $\{I_1, I_2, \ldots I_k\}$ is a k-partition of $N = \{1, 2, \ldots, n\}$ if:

- $I_1 \cup I_2 \ldots \cup I_k = N$

- $I_i \cap I_j = \phi \qquad i \neq 1 \quad i,j = 1 \dots k$

- $I_i \neq \phi \qquad i = 1 \dots k$

We call $\{I_1, I_2 \dots I_k\}$ an <u>independent k-partition</u> if the $I_{i's}$ are independent sets.

If we let:

$Y_{n,k}$ = number of independent k-partitions of $\{1, 2, \dots n\}$(a random variable).

$Z_{n,k}$ = number of k-partitions of $\{1, 2, \dots n\}$

Then clearly:

$P_r \{X(w_n) \leq k\} = P_r \{ \exists$ an independent k-partition of $\{1, 2, \dots n\}\}$

$$= P_r \{Y_{n,k} > 0\} \leq E \{Y_{n,k}\}$$

An explicit formula can be obtained for $E(Y_{n,k})$, and a value for k can be selected so that $\lim_{n\to\infty} E(Y_{n,k}) = 0$. The following result can be shown [3].

<u>Theorem 9</u>. With probability going to 1 as $n\to\infty$, it holds that $X(w_n) \geq k^*$ where $k^* = \text{Max} \{k^k < e^c\}$.

This result can be used to estimate the error of the sequential coloring algorithm. Also, the following inequality holds for $X(w_n)$:
$k^* \leq X(w_n) \leq X^* + 3$

Using the formulas for $X^*$ and $k^*$, one can observe that:
$\lim_{c\to\infty} \frac{X^* + 3}{k^*} = 2$, a result that was proposed previously by Erdos [1].

## 4. CONCLUSIONS AND FURTHER RESEARCH

The probabilistic analysis presented in this paper can be extended to more sophisticated algorithms and to new problems. For instance, one can consider the following 2-machine flowshop problem: n jobs are available at t = 0 on the first machine, their processing times being $P_1$, $P_2$ ... $P_n$. Upon completion of service on machine 1, jobs must undergo processing on machine 2, with processing times $Q_1$, $Q_2$, ... $Q_n$. The Pi's are i.i.d. randoms variables and so are the Qi's. Total flow time must be minimized. Asymptotically optimal algorithms for the case of exponential distributions of the processing times will be presented on a future paper.

For the chromatic number problem, a more sophisticated algorithm can be analyzed. Suppose that we examine the nodes sequentially and include a node in the solution if it has at most $(k_1-1)$ edges to the previously included ones. It is easy to see that such a set can be colored with at most $k_1$ colors. If the nodes in the set are removed, a new random graph is obtained and a node set can be identified that can be colored with $k_2$ colors, using the same procedure as before. The algorithm continues until all nodes are included in such sets, and a coloring of the graph is produced using at most $k_1 + k_2 + \ldots +$ etc. colors. A probabilistic analysis of this algorithm can be performed.

If the integers $k_1$, $k_2$ ... etc. are chosen appropiately, a feasible solution can be obtained that compares favorably with the bound on the optimal solution.

## REFERENCES


1. Erdos, P., Spencer, J. Probabilistic Methods in Combinatorics. Academic Press, 1974.

2. Gazmuri, P. "Probabilistic Analysis of Packing, Coloring and Scheduling Problems". Ph.D. Thesis, University of California, Berkeley, 1980.

3. Gazmuri, P. "Chromatic Number on Sparse Graphs". Working Paper N°84/06. Depto. de Ingeniería de Sistemas. Pontificia Universidad Católica de Chile.

4. Gazmuri, P. "Probabilistic Analysis of a Machine Scheduling Problem", Mathematics of Operations Research Vol. 10, N°2, 1985.

5. Gazmuri, P. "Some new results on probabilistic analysis of machine scheduling problems". Working Paper N°85/01. Depto. de Ingeniería de Sistemas. Pontificia Universidad Católica de Chile. Submitted for publication in Operations Research.

6. Gazmuri, P. "Independent sets in Random Sparse Graphs". Networks, Vol 14, N°3, 1984.

7. Graham, R.L., Lawler, E.L., Lenstra, J.K., Rinnoy Kan, A.H.G. "Optimization and Aproximation in Deterministic Sequencing and Scheduling: a survey". Annals of Discrete Mathematics Vol 5, 1979.

8. Grimmett, G.R. McDiarmid, J.H. "On coloring random graphs". Math. Proc. Camb. Phil. Society Vol 77, 1975.

9. Karp, R.M. "Probabilistic Analysis of Partitioning algorithms for the Travelling-Salesman Problem in the plane". Mathematics of Operations Research. Vol 2, N°3, 1977.

10. Karp, R.M. "The Probabilistic Analysis of some combinatorial search algorithms". Memorandum N°ERL-M581, 1976. Electronics Research Laboratory, University of California, Berkeley.

11. Loulou, R. "Tight Bounds and Probabilistic Analysis of two heuristics for Parallel Processor Scheduling" Mathematics of Operations Research Vol 9, N°1, 1984.

12. Pinedo, M., Schrage, L. "Stochastic Shop Scheduling: A survey" in Deterministic and stochastic Scheduling. Nato Advanced Study Institutes Series. D. Reidel Publishing Company. 1981.

13. Wiss, G. "Multiserver Stochastic Scheduling" in Deterministic and Stochastic Scheduling. Nato Advanced Study Institute Series. D. Reidel Publishing Company, 1981.

# THE ELLIPSOID METHOD AND ITS PREDECESSORS

Jean-Louis Goffin
Faculty of Management, McGill University
1001 Sherbrooke Street West
Montreal, P.Q. H3A 1G5
Canada

## 1. INTRODUCTION

The relaxation method of Agmon[1], Motzkin and Schoenberg[14] was one of the early proposals made to solve linear programs. It was shown very quickly by Hoffman et al.[12] not to be competitive with the simplex method of George Dantzig. Neither of these methods turned out to be polynomial according to the theory of computational complexity developed in the late sixties.

The relaxation method, which is designed to solve systems of linear inequalities, was adapted to the problem of minimizing a general convex function in works by Shor going back to 1961; it was proposed later, and independently, by Held and Karp[10] and Held, Wolfe and Crowder[11] and became known as subgradient optimization. Subgradient optimization, like the relaxation method, worked well on some problems, but very poorly on many others; the rate of convergence depends critically on the fact that the function, or the system of inequalities, are well or poorly conditioned (this is a property which depends upon angles, which are defined in analogous ways for both methods).

In 1969-1970, Shor proposed modifications to the subgradient algorithm which use a variable metric, as is done by all quasi-Newton methods; two methods were proposed, one where the space is dilated in the direction of the subgradient (this became known later as the ellipsoid method) and a second where the dilatation takes place along the difference of two successive gradients (this was given the name of r-algorithms). Both of these methods were used by Shor and his colleagues, and excellent computational results were achieved. Most of Shor's work (theory and computational experiments) up to 1979 is very well described in his book "Minimization Methods for Non-Differentiable Functions"[15].

The ellipsoid method is a variant of the method of dilatation in the direction of a subgradient in which a specific choice of two scalar parameters was given by Yudin and Nemirovskii[16]; this choice was used by Khacian[13], to show that the linear programming problem can be solved in polynomial time. This specific choice of parameters made the method quite slow, especially in medium or large scale linear problems. Positive results have been reported by Ecker and Kupferschmid (see[3], for instance) on smaller and nonlinear problems.

We have been able to show ([8],[9]) that the deepest cut variant of the ellipsoid method applied to a system of linear inequalities will compute an optimal, or almost optimal, metric; formally, the ellipsoid method with deepest, or least shallow, cuts is a polynomial (time and space) method which finds an ellipsoid representable by polynomial space integers such that the maximal ellipsoidal distance relaxation method, using this ellipsoid, is polynomial. Intuitively, the ellipsoid method reshapes the problem so that the angles of the solution set are not small (i.e. bounded below by the inverse of the dimension of the space).

Similar results do not extend, in general, to subgradient optimization unless additional conditions are put on the function, such as symmetry.

Experimental results on a Lagrangean relaxation formulation of nonlinear multicommodity flow problems are presented; the results are somewhat mixed: the computational times are very good when the method works, but it is quite difficult to tune parameters correctly to make the method reliable, unless, it seems, the Lagrangean relaxation is unconstrained (which corresponds to the case of an acyclic network).

## 2. SUBGRADIENT OPTIMIZATION AND THE ELLIPSOID METHOD

Let $f(x)$ be a convex function defined on $R^n$, which is assumed to have a minimum $f^*$ attained on a set of minimum points, $P = \{x \in R^n : f(x) = f^*\}$. If f is polyhedral then $f(x) = \text{Max}\,\{e_i^t(Ax-b) : i \in M\}$, where M is a finite index set, $e_i$ is a unit vector with a 1 on position i, A is a matrix and b a vector of appropriate dimensions.

For every value of x, an oracle (i.e. a subproblem) returns the value $f(x)$ and a subgradient $g(x)$ belonging to $\partial f(x)$, the subdifferential of f at x; the pair $(f(x),g(x))$ defines a supporting plane (or cutting plane) to the epigraph of f. If f is differentiable, the $\partial f(x) = \{f'(x)\}$, and thus $g(x)$ is the gradient of f at x.

Subgradient optimization will move in the direction of the negative of a subgradient by a step size that cannot be based on a line search, as the function f need not decrease in the direction $-g(x)$. One function which decreases in the direction $-g(x)$ is the Euclidean distance to the optimal set P, $d(x,P) = \text{Max}\{\|x-x^*\| : x^* \in P\}$, which is not computable, but can be used to prove convergence results for open loop, or off line, step sizes.

The subgradient algorithm as defined by:

$$x_+ = x - sg(x)/\|g(x)\|,$$

where + will be used to denote the <u>next</u> iterate in the sequence $\{x_q : q = 0,1,\ldots\}$. The use of a normalization of $g(x)$ by its $L_2$ norm is required in order to get a convergence theory, involving rates of convergence, with similar step size rules for differentiable as well as nondifferentiable functions. The theory of convergence depends on condition numbers, or on the angles between the direction chosen $(-g(x))$ and the ideal direction $(x^*(x) - x)$, where $x^*(x)$ is the Euclidean projection of x on the minimal set P.

Definition 2.1.

The condition number $\mu = \sin \eta = \cos(\pi/2 - \eta)$ of a convex function is:

$$\mu = \inf_{x \notin P} \max_{g \in \partial f(x)} (g, x - x^*(x))/(\|g\| \, \|x-x^*(x)\|).$$

For every convex, or even quasiconvex, function $\mu \in [0,1]$; note that the gradient of $d(x,P)$ is $(x-x^*(x))/\|x-x^*(x)\|$, so that $-g(x)$ is a descent direction for $d(x,P)$.

Theorem 2.2.

Let f be a convex function defined on $R^n$; assume that the angle $\eta$ satisfies $0 < \eta \leq \pi/2$ and define

$\alpha = \cos^2 \eta \quad \delta = \sin \eta \qquad$ if $\eta \in (0,\pi/4]$

$\alpha = (1/2 \sin \eta)^2 \quad \delta = 1/2 \sin \eta \quad$ if $\eta \in [\pi/4,\pi/2]$

If we choose in the subgradient algorithm the step size to be $s_q = d\,\delta\alpha^{q/2}$ where d is an upper bound on $d(x_o,P)$, then one has:

$$d(x_q,P) \leq d\,\alpha^{q/2} \qquad \text{for all } q;$$

and if f is Lipschitz with constant K, one has:

$$f(x_q) - f^* \leq Kd\,\alpha^{q/2} \text{ for all } q.$$

A proof of this may be found in Shor's book[15] or in [5]; it is based upon the idea that for the values of $\alpha,\delta$ and d given, and for all q, a sphere of radius $d\,\alpha^{q/2}$ centered at $x_q$ has a nonempty intersection with P.

It follows that the number of iterations needed to achieve a precision $\Delta$, in terms of the function values, has an upper bound given by (assuming $\eta \leq \pi/4$)

$$q \leq 2(\sin \eta)^{-2} \ln(Kd/\Delta).$$

If the angle $\eta$ is small, which happens often, but far from always, in applications, the method is so slow as to be quite useless. If $\eta = \pi/2$, f is perfectly conditioned, and the achievable rate of convergence is 1/2, the rate of the bisection method.

The classical way to improve convergence of a first-order method in the case of a poorly conditioned function is to apply the original method in a transformed space, where hopefully the transformed function is better conditioned; the most classical transformation is an affine transformation (note that projective transformations have become very popular too) $x = Ty$, where T is a square, nonsingular, matrix. A fully implementable algorithm (like Newton or quasi-Newton Methods) would iterate on the pair $(x,T)$, but for now we will assume T fixed.

Define $\varphi(y) = f(Ty)$, then it can be shown that:

$$\partial\varphi = T^t\, \partial f$$

$$\varphi^* = f^*$$

$$\{y \in R^n : \varphi(y) = \varphi^*\} = T^{-1}P;$$

the subgradient method applied to $\varphi$ is given by:

$$y_+ = y - s\,\gamma(y)/\|\gamma(y)\| \quad , \quad \text{where} \quad \gamma(y) \in \partial\varphi(y).$$

The rate of convergence will depend on a condition number, or angle, defined for the function $\varphi$, which should be better than those associated with f; for instance if f is convex and $C^2$, the optimal choice for T is the square root of the inverse Hessian of f at its minimum.

Theorem 2.2 applies, with obvious changes to the sequence $y_q$: $q = 0,1,2,\ldots$ defined by the subgradient algorithm applied to the function $\varphi$ in y-space. This gives rises to a new algorithm (essentially the ellipsoid method with a fixed ellipsoid) in x-space by defining

$$\{x_q\} = \{Ty_q\}.$$

Algorithm 2.3. ("variable" metric subgradient optimization)

$$x_+ = Ty_+ = Ty - s\,T\gamma(y)/\|\gamma(y)\|$$

$$= x - sHg(x)/(g^t(x)\,H\,g(x))^{1/2}$$

with $g(x) \in \partial f(x)$ and $H = TT^t$.

The matrix H is positive definite, and in the classical $C^2$ case, it should be, or approximate, the inverse Hessian of f at its minimum; algorithm 2.3 can then be interpreted as a conceptual version of Newton's method.

A convergence theory for algorithm 2.3 simply translates theorem 2.2 applied in y-space to the space of the original variable x. The $L_2$ norm in y-space translates in x-space as the ellipsoidal norm:

$$(x^t H^{-1} x)^{1/2} = \|T^{-1}x\| = \|y\|.$$

The projection of x on P according to this norm will be denoted by $x_H^*(x)$ and the ellipsoidal distance from x to P by $d_H(x,P) = \mathrm{Min}\{[(x'-x)^t\, H^{-1}(x'-x)]^{1/2};\ x' \in P\}$; the minimizer is $x_H^*(x)$.

Definition 2.4.

The condition number $\mu_H = \sin \eta_H$ of the function $\varphi$ is:

$$\mu_H = \inf_{y \notin T^{-1}P}\ \min_{\gamma \in \partial\varphi(y)} (\gamma, y - y^*(y))/(\|\gamma\|\, d(y, T^{-1}P));$$

expressed in terms of the function f it is:

$$\mu_H = \inf_{x \notin P}\ \min_{g \in \partial f(x)} (g, x - x_H^*(x))/((g^t Hg)^{1/2}\, d_H(x,P)).$$

Theorem 2.5.

Let f be a convex function defined on $R^n$, $H = TT^t$ a $n \times n$ positive definite matrix, assume that the angle $\eta_H$ satisfies $0 < \eta_H \leq \pi/2$, and define

$$\alpha_H = \cos^2 \eta_H \qquad \delta_H = \sin \eta_H \qquad \text{if} \quad \eta_H \in (0,\pi/4]$$

$$\alpha_H = (1/2 \sin \eta_H)^2 \qquad \delta_H = 1/2 \sin \eta_H \quad \text{if} \quad \eta_H \in [\pi/4,\pi/2] \quad .$$

If we choose in algorithm 2.3 the step size to be $s_q = d_H \delta_H \alpha_H^{q/2}$ where $d_H$ in an upper bound on $d_H(x_o,P)$, then one has:

$d_H(x_q,P) \leq d_H \alpha_H^{q/2}$ for all q; and if f is Lipschitz with constant K, one has:

$f(x_q) - f^* \leq K \;\Lambda_H^{1/2}\; d_H \alpha_H^{q/2}$ for all q, where $\Lambda_H$ is the largest eigenvalue of H.

Proof:

The proof is an exact copy of the proof of Theorem 2.2 with the ellipsoidal norm $(x^t H^{-1} x)^{1/2}$ replacing the Euclidean norm. For the last statement of the theorem:

$$\begin{aligned} f(x_q) - f^* &\leq K \;\| x_q - x^*_H(x_q) \| \\ &= K \,\| H^{1/2} H^{-1/2} (x_q - x^*_H(x_q)) \| \\ &\leq K \,\| H^{1/2} \|\, d_H(x_q,P) \\ &\leq K \;\Lambda_H^{1/2} \; d_H \;\alpha^{q/2} \quad . \end{aligned}$$

The proof is based upon the idea that the ellipsoid

$$E_q = \{x' \in R^n : (x' - x_q)^t H^{-1} (x' - x_q) \leq d_H^2 \;\alpha^q\}$$

has a nonempty intersection with the minimal set P, and this for all q.

It follows that the number of iterations needed to achieve a precision $\Delta$, in terms of the function values, has an upper bound given by (assuming $\eta_H \leq \pi/4$)

$$q \leq 2(\sin \eta_H)^{-2} \;\ell n(K \;\Lambda_H^{1/2}\; d_H/\Delta).$$

One can expect that if H as well chosen $\sin \eta_H$ will be greater than $\sin \eta$, and thus the rate of convergence of algorithm 2.3 would be better than that of subgradient optimization.

By analogy with results proved in [8] and [9] about the maximal distance relaxation method of Agmon, Motzkin and Schoenberg, one could be tempted to conjecture that for every f a positive definite matrix $H = TT^t$ exists and satisfies $\sin \eta_H \geq 1/n$, and furthermore the ellipsoid method would, in some sense, approximate such an H. The first part of this conjecture turns out not to be true without additional assumptions (symmetry or approximate symmetry); the second part (i.e. the ellipsoid method approximates such an H) has been shown to be true in the special case where f is the $L_\infty$ norm of a full rank system of linear equalities, and the use of equalities is based upon an a priori cyclical order [6].

Algorithm 2.7. : The maximal distance relaxation method.

Problem: find $x \in P = \{x \in R^n : Ax \leq b\}$

Method : project the current iterate on the most distant violated inequality

Analogy with subgradient optimization:

$$f(x) = \text{Max}\ \{0, \underset{i \in I}{\text{Max}}\ e_i^t(Ax - b)/(e_i^t AA^t e_i)^{1/2}\}$$

$$x_+ = x - f(x)\ g(x)/\|g(x)\| \quad g(x) \in \partial f(x) \ .$$

Angle: $\eta$ as in definition 2.1 (note: $\mu$ is Agmon's $\mu$)

Convergence: rate of convergence: $\cos \eta$; number of iterations proportional to $(\sin \eta)^{-2}$.

Because of the normalization (by the norms of the rows of A) implicit in the definition of f, the maximal distance applied in y-space ($x = Ty$, $H = TT^t$) will not use the function $\varphi(y) = f(Ty)$ but the function

$$\bar{\varphi}(y) = \text{Max}\ \{0, \underset{i \in I}{\text{Max}}\ e_i^t(ATy - b)/\|T^t A^t e_i\|\}\ ;$$

in the original space the function

$\bar{f}(x) = \bar{\varphi}(T^{-1}x)$ will be used. The condition number associated to the function $\bar{\varphi}$ (see definition 2.4) will be denoted by $\bar{\mu}_H = \sin \bar{\eta}_H$ .

Algorithm 2.8.: The maximal ellipsoidal ($H = TT^t$) distance relaxation method.

$$\bar{f}(x) = \text{Max}\ \{0, \underset{i \in I}{\text{Max}}\ e_i^t(Ax - b)/(e_i^t AHA^t e_i)^{1/2}\}$$

$$x_+ = x - \bar{f}(x)H\bar{g}(x)/(\bar{g}^t(x)H\bar{g}(x))^{1/2}$$

where $\bar{g}(x) \in \partial\bar{f}(x)$

Convergence: rate of convergence: $\cos \bar{\eta}_H$; number of iterations proportional to $(\sin \bar{\eta}_H)^{-2}$ .

The deepest cut ellipsoid method (see [ 9 ] for a complete description) is a version of algorithm 2.8 which iterates on x and on H. The following theorem rephrases results proved in [8] and [9].

Theorem 2.9.

Let $Ax \leq b$ be a system of linear inequalities with nonempty, full bodied, and compact solution set P. Then there exists matrices $H = TT^t$ such that $\sin \bar{\eta}_H \geq 1/n$; such matrices may be given by the largest ellipsoid contained in P, or the smallest ellipsoid containing P.

Every limit point $H^*$ of the sequence of matrices generated by the deepest cut ellipsoid method satisfies $\sin \bar{\eta}_{H^*} \geq 1/n$.

The key to the proof of Theorem 2.9 is a classical result due to Fritz John which says that the affine excentricity of any compact convex set in $R^n$ is at most n.

The ellipsoid method, applied to the problem of minimizing a convex function f, will iterate, like most quasi-Newton methods, on x but also on H. A proper view of these methods is that they search not only for an optimal point $x^*$, but also for an optimal metric $H^*$ (if such an optimal metric exists).

Algorithm 2.10: A class of ellipsoid (variable metric subgradient methods: (Shor[15]).

$$x_+ = x - s\,Hg/(g^tHg)^{1/2}$$

$$H_+ = H - \beta Hg\,g^tH/(g^tHg) \qquad g \in \partial f(x)\ .$$

Theorem 2.11. (The ellipsoid method)

Let f be a convex function defined on $R^n$. If we define $\alpha = n^2/(n^2-1)$, $\beta = 2/(n+1)$, $\delta = 1/(n+1)$ and $s_q = d_{H_o}\,\delta\alpha^{q/2}$ where $d_{H_o}$ is an upper bound on $d_{H_o}(x_o,P)$, then

$$d_{H_q}(x_q,P) \le d_{H_o}\,\alpha^{q/2} \quad \text{for all } q$$

and if f is Lipschitz with constant K

$$\min_{j=o,\dots,q} f(x_j) \le f^* + [\alpha(1-\beta)^{1/n}]^{q/2}\,K\,d_{H_o}\,\Lambda_{H_o}^{1/2}\ .$$

Furthermore if $\Lambda_{H_q} \le K_1(1-\beta)^{q/n}$, which imply that the condition number of $H_q$ is bounded, then

$$f(x_q) \le f^* + K K_1^{1/2} d_{H_o} [\alpha(1-\beta)^{1/n}]^{q/2} \quad \text{for all } q.$$

The proof (see [ 7 ]) is based upon the idea that the ellipsoid

$$E_q = \{x' \in R^n : (x'-x_q)^t H_q^{-1}(x'-x_q) \le d_{H_o}^2\,\alpha^q\}$$

has a nonempty intersection with the minimal set P, and this for all q. The assumption that $\Lambda_{H_q} \le K_1(1-\beta)^{q/n}$, together with the fact that $\det H_q = (1-\beta)^q \det H_o$, implies that all eigenvalues of $H_q$ are bounded, above and below, by geometric series of ratio $(1-\beta)^{q/n}$; and therefore $\{(1-\beta)^{-q/n}H_q\}$ would have limit points. It is well known that $\alpha(1-\beta)^{1/n} \le e^{-1/n(n+1)}$, and thus that the number of iterations needed to reach an accuracy of $\Delta$, in terms of the recorded function values, is:

$$q \le 2n(n+1)\ln(K\,d_{H_o}\,\Lambda_{H_o}^{1/2}/\Delta).$$

It is not true, in general (there are counterexamples), that the eigenvalues of $\{(1-\beta)^{q/n}H_q\}$ are bounded. It is also not true that there exist matrices H such that $\sin \eta_H \ge 1/n$; so a theorem like theorem 2.9 does not seem to hold in the case of an optimization problem. A rationale for this discrepancy may be that an optimization problem has one more variable, the objective, than that of solving linear inequalities; one should thus look, not at affine transformations defined on $R^n$, but either at affine transformations defined on $R^{n+1}$ or at projective transformations defined on $R^n$.

It is possible to show results as in the first part of theorem 2.9 under additional conditions like symmetry (or approximate symmetry); a proof of something similar to the second part of theorem 2.9 (convergence results on $(1-\beta)^{q/n}H_q$) should be feasible, but this is still a conjecture.

Theorem 2.12.

Let f be a convex function defined on $R^n$. Assume that f is even and positively homogeneous, i.e. $f(tx) = |t|^a f(x)$ for all $t \in R$, and some $a \geq 1$; then there exist matrices $H = TT^t$ such that $\sin \eta_H \geq 1/n^{1/2}$.

Proof:

Clearly $f^* = 0$, $x^* = 0$ and $X = \{x \in R^n : f(x) \leq 1\}$ is a convex set symmetrical with respect to the origin; we will assume X bounded (i.e. $f(x) = 0 \Rightarrow x = 0$). The centre of gravity of X is the optimum point $x^* = 0$, and all level sets of f are homothetic to X.

Let $E = \{x \in R^n : x^t H^{-1} x \leq 1\}$ be the largest volume ellipsoid included in X (one could work with the smallest ellipsoid containing X); by symmetry, the center of E is the origin (note: this is where symmetry is critical). It is well known in geometry that

$$E \subset X \subset n^{1/2} E.$$

If we let $H = TT^t$, $\varphi(y) = f(Ty)$ and $Y = T^{-1}X$, then, with S denoting the unit ball,

$$S \subset T^{-1}X \subset n^{1/2} S .$$

this immediately implies that $\sin \eta_H \geq 1/n^{1/2}$. □

The key technical difference between this theorem and theorem 2.9 is that here the ellipsoid is restricted to have its center at $x^*$, while this was not required in the case of a system of linear inequalties (where a renormalization of the constraints was allowed).

Examples

1. $f = \frac{1}{2} x^t Qx$ Q positive definite, symmetric

   define $H = Q^{-1}$, $x = Q^{-1/2}y$, then

   $\varphi = \frac{1}{2} \| y \|^2$ and $\sin \eta_H = 1$.

2. $f = \| A x \|_\infty$ ($A \in R^{n,n}$, non singular)

   define $H = (A^tA)^{-1}$ $x = A^{-1}y$

   then $\varphi = \| y \|_\infty$ and $\sin \eta_H = 1/n^{1/2}$ .

3. $f = \| A x \|_1$ ($A \in R^{n,n}$, non singular)

   define $H = (A^tA)^{-1}$ $x = A^{-1}y$

   then $\varphi = \| y \|_1$ and $\sin \eta_H = 1/n^{1/2}$ .

4. $f = \| Ax \|$ $(A \in R^{n,n},$ non singular)

define $H = (A^t A)^{-1}$ $x = A^{-1} y$

then $\varphi = \| y \|$ and $\sin \eta_H = 1$ .

All these examples are symmetrical functions, and theorem 2.12 breaks down as soon as X as a simplex, for instance, without additional conditions (such as: 0 is the centre of gravity of X).

## 3. Computational results on nonlinear multicommodity flow problems.

The objective used in most experiments represents the sum of the queuing delays of total arc flows $y_a$ in arcs a of capacity $c_a$, i.e. $\sum_a y_a/(c_a - y_a)$; quadratic objectives were also used, but were not quite as difficult as the queuing delay objectives.

The problem is:

$$\text{Min} \quad \sum_a y_a/(c_a - y_a)$$

$$\text{s.t.} \quad E x^k = f^k \qquad x^k \geq 0 \quad k \in K$$

$$\sum_{k \in K} x^k = (\text{or} \leq)\ y\ , \qquad y \geq 0$$

where K is usually identified with the number of distinct origins, E is the node-arc incidence matrix, $f^k$ are the requirments of commodity k, and $x_a^k$ is the flow of commodity k in arc a. The number of arcs is n.

By introducing multipliers $\lambda = (\lambda_a)$ on the coupling constraints, one defines a partial Lagrangean:

$$L(\lambda) = \min_{y,x^k} \{ \sum_a (y_a/(c_a - y_a) - \lambda_a y_a) + \sum_{k \in K} (\lambda, x^k) : Ex^k = f^k, x^k \geq 0, k \in K;\ y \geq 0 \}$$

$$= \sum_a f_a^*(\lambda_a) + \sum_{k \in K} \text{Min} \{ (\lambda, x^k) : Ex^k = f^k,\ x^k \geq 0 \}$$

where
$$f_a^*(\lambda_a) = \begin{cases} -(1 - \sqrt{\lambda_a c_a})^2 & \text{if } \lambda_a \geq 1/c_a \\ 0 & \text{otherwise} \end{cases}$$

$$y_a(\lambda) = \begin{cases} c_a - \sqrt{c_a/\lambda_a} & \text{if } \lambda_a \geq 1/c_a \\ 0 & \text{otherwise} \end{cases}$$

$x^k(\lambda)$ is computed by using a $\lambda$-shortest path between the origin and destinations for commodity k (assuming no $\lambda$-negative cycles).

The dual problem may be formulated either as:

$$\max_\lambda \{ L(\lambda) : \text{ there are no } \lambda\text{-negative cycles } \}$$

or

$$\max_\lambda \{ L(\lambda) : \lambda \geq 0 \}.$$

The function $L(\lambda)$ is the sum of a $C^1$ function, which is also piecewise $C^2$, and a piecewise linear function; it is concave, and $L(\lambda) = -\infty$ when there are $\lambda$-negative cycles.

A subgradient to $L(\lambda)$ at $\lambda$, if there are no $\lambda$-negative cycles, is:

$$g = -y(\lambda) + \sum_{k\in K} x^k(\lambda);$$

the subdifferential $\partial L(\lambda)$ is $\partial L(\lambda) = -y(\lambda) + \sum_{k\in K} X^k(\lambda)$, where $X^k(\lambda)$ is the set of all solutions $x^k(\lambda)$ (i.e. the convex hull of all optimal extreme points, or tree solutions) to $\text{Min}\,\{(\lambda,x^k): Ex^k = f^k,\ x^k \geq 0\}$.

If the condition $\lambda \geq 0$ is enforced within the master problem (for instance, by projecting on the set $\{\lambda \geq 0\}$), then the subproblems will not have any negative cycles. If the condition $\lambda \geq 0$ is not enforced, the subproblems should return one, or many $\lambda$-negative cycles; any positive linear combination of indicator vectors of $\lambda$-negative cycles could be returned as the normal vector to a hyperplane separating the current point $\lambda$ from the set

$$N = \{\lambda : \text{there are no } \lambda\text{-negative cycles}\}.$$

The method that we used for computing $\lambda$-shortest paths or $\lambda$-negative cycles is a slight modification of the Floyd-Warshall algorithm, in which the procedure is not stopped when a negative cycle is detected, but runs to the end of the three Do loops; the node pointers refer to the predecessor on the path or cycle. Our expectation is that $\lambda$-negative cycles, if they exist, will be found for more than one origin destination pair, and thus provide us with a better separation between $\lambda$ and N.

Negative cycles are identified when backtracking from a destination to an origin, and the flow around the cycle is put to the origin-destination flow multiplied by a large number (penalty). The subgradient returned by the subproblem will then be, as above, $g = -y(\lambda) + \sum_{k\in K} x^k(\lambda)$.

The following choices were made for the computational experiments:

1. A heuristic choice of the parameters $\alpha,\beta$ and $\delta$:
   a) ellipsoid method: $\alpha=(1/2)^{1/n}$, $\beta = 1/2$, $\delta = 1$;
   b) subgradient optimization: $\alpha=(1/4)^{1/n}$, $\delta = 1$; this corresponds to the idea of halving the step-size every n iterations proposed by Held, Wolfe and Crowder [11].

This choice of parameters implies that, if convergence occurs, the number of iterations will be a multiple of n (and not $n^2$, as in the choice of $\alpha,\beta,\delta$ which guarantee convergence).

2. The starting point ($\lambda^o$) and an upper bound (d) on the distance between $\lambda^o$ and an optimal solution ($\lambda^*$): note that the optimal arc flows $y^* = y(\lambda^*)$ are unique (the primal objective is strictly convex), and that for all arcs

a which are used (i.e. $y^*_a > 0$) one has

$$\lambda^*_a = \frac{1}{c_a} \frac{1}{(1-\rho^*_a)^2}, \quad \text{where } \rho^*_a = y^*_a/c_a \text{ is the traffic intensity}$$

in arc a at the optimum; so we tried to guess an upper bound $\rho$ on $\operatorname{Max}_a \rho^*_a$, and used

$$\lambda^o_a = \frac{1}{c_a} \frac{1}{(1-\rho)^2}$$

and $\quad d \approx (\sum_a c_a^{-2})^{1/2} (1-\rho)^{-2}$ (where $\approx$ means: of the order of).

3. The constraints, i.e. $\lambda \geq 0$ or $\lambda \in N = \{\lambda$:there are no $\lambda$-negative cycles$\}$; two options were tried:

   a) Enforce $\lambda \geq 0$ by projecting orthogonally on $\lambda \geq 0$, after each step of the ellipsoid method, or of subgradient optimization ;

   b) Use a separator between $\lambda$ and N as a subgradient to be used by either method;

   a third option which may be sensible would be, in the case of the ellipsoid method, to project on $\lambda \geq 0$ using the metric given by the current ellipsoid.

4. The test problems; many were used, but we shall report on only two, as the other test problems behaved somewhat similarly:

   a) a small problem described in [4], with 22 acres, 14 nodes, 23 O-D pairs, 5 commodities and <u>no</u> cycles (i.e. an acyclic graph); the O-D flow from node 5 to node 8 was set at 6.75 (versus .75 on [4]), so as to get a high-traffic intensity in two arcs ($\rho^* = \operatorname{Max}_a \rho_a = 98.4\%$), and optimal Lagrange multipliers which range between .03 and 364. The primal formulation has 93 rows, 132 variables (22 of them nonlinear), and 352 non zero elements; the optimal value of the objective is 103.41.

   b) a large problem (a copy of the ARPANET topology as of August 1978) described in [2] with a small correction (in [2], there are 2 nodes 19, and no node 49; the bottom node 19 was named node 49); it does have 61 nodes, 148 arcs, 122 O-D pairs, 61 commodities. The traffic intensity at the optimum is 87%, while optimal Lagrange multipliers range between .8 and 12. When an arc appears between two nodes, the reverse one is also present, and therefore there are many cycles.
   The primal formulation has 3870 rows, 9176 variables (148 nonlinear variables) and 27232 nonzero elements; the optimal value of the objective is 151.93.

5. Implementation.

The ellipsoid matrix H was not used explicitly, but a factor J of H (i.e. $H = JJ^t$) was kept and updated.

The termination criterion was based both on a maximum limit on the number of iterations, and on the size of $\|J_q^t g_q\| \alpha^{q/2}$ (which is an upper bound on $f(x_q)-f^*$, when the parameters are those given by Yudin and Nemirovskii, but turns out to be an upper bound on $f(x_q)-f^*$ for other choices of parameters, provided that they lead to convergence).

The computer used was McGill University's Amdahl 5850, and the FORTRAN compiler used when running MINOS was FORTRAN G1, while when running the ellipsoid method or subgradient optimization it was FORTRAN VS (level = 77, optimization = 3). The smaller problem ran easily on an IBM PC with an 8087 mathematics coprocessor. All work on the ellipsoid method was done in single precision.

6. Problem a) (see 4a) above)

MINOS reached $f^* = 103.412$, after 69 iterations, 88 function and gradient calls, and 0.91 secs of CPU time. An estimate of the condition number of the Hessian at the optimum is $3.2\ 10^5$.

The same problem formulated with 23 commodities (rather than 5) required MINOS to work for 11.42 secs of CPU time.

Using the choice of parameters given in 1a) and 1b) above, $\rho = .99$ and $d = 3162$, the following implementations failed :

- ellipsoid method with projection on $\{\lambda: \lambda \geq 0\}$
- subgradient optimization
- subgradient optimization with projection on $\{\lambda : \lambda \geq 0\}$;

the only implementation which reached the optimum was

- the ellipsoid method:

it reached the optimum $f^* = 103.471$ in 500 iterations (stopping criterion) and a CPU time of 2.42 secs. Note that because this is a dual approach the objective should be <u>less</u> than that given by MINOS (a primal approach), but rounding errors are present in a significant manner. Negative cycles were never generated, as the graph is acylic.

7. Problem b) (see 4 b) above)

MINOS reached $f^* = 151.927$ after 14909 iterations (12846 linear iterations were needed to find a feasible basis), after 2777 function and gradient calls, and 58 mins 04.93 secs. An estimate of the condition number of the Hessian is 150.

Using the choice of parameters given in 1 a) and 1 b) above, $\rho = .86$, $d = 3.16$, the following implementation failed:

-the ellipsoid method (with no projection on $\{ \lambda: \lambda \geq 0\}$ and generation of negative cycles); the following implementations reached the optimum:

- the ellipsoid method with projection on $\{ \lambda : \lambda \geq 0\}$: $f^* = 151.81$, 1000 iterations, 4 mins 49.93 secs of CPU time

- subgradient optimization (with no projection on $\{\lambda: \lambda \geq 0\}$ and generation of negative cycles): $f^* = 151.228$, 1000 iterations, 4 mins 37.68 secs of CPU time

- subgradient optimization (with projection on $\{\lambda: \lambda \geq 0\}$): $f^* = 151.26$, 1000 iterations,4 mins 39.78 secs.

Other runs were performed for different values of the main tuning parameters $(\alpha,d,\rho)$ but most of them did not affect the qualitative conclusion that the ellipsoid method with no projection on the positive orthant and with generation of negative cycles failed, while the three other implementations were somewhat robust. For instance changing $\alpha$ from $(1/2)^{1/n} = .9953$ to $\alpha = .9$ gave similar results (except that the CPU time was cut to 1 minute, or less); this indicates that this problem is rather well conditioned.

In the case of the ellipsoid method with no projection on the positive orthant and with generation of negative cycles, the values of $\alpha,\beta,\delta$ giving theoretical convergence( $\alpha = 1.000046$ $\beta = .0135$ $\delta = .00067$) were tried, and convergence to within 5% of the optimal value took place rather quickly ($L(\lambda^o) = -6407$, $L(\lambda^{322}) = 144.11$), while the ensuing iterations improved the objective at an extremely slow pace ($L(\lambda^{10000}) = 149.85$).

It is probably worth noting that one iteration of MINÖS and one iteration of the ellipsoid method took almost exactly the same time (1/4 sec ).

## 4. Conclusions

Convex functions which are symmetrical and positively homogeneous, with respect to their minimum point, can be transformed by affine transformations into convex functions for which subgradient optimization converges at a linear rate which depends only upon the dimension of the space. By analogy with theoretical results showing, in particular instances ([6],[9]), that the ellipsoid method would find such an affine transformation, and with experimental results ([5]) indicating a possible rate of convergence of $1 - O(n^{-1})$ rather than the theoretical value of $1-O(n^{-2})$ (the discrepancy coming, quite possibly, from an acceleration phenomenon typical of relaxation methods), we expected that the ellipsoid method, with a heuristic choice of parameters, would lead to convergence in a number of iterations proportional to n(the dimension of the space).

Experimental results on a Lagrangean relaxation formulation of nonlinear multicomodity flow problems are somewhat mixed.

On a small, poorly conditioned, unconstrained (i.e. an acyclic network) problem, computational results confirmed our expectations.

A large, well conditioned, problem with many constraints (i.e. a graph with many cycles) made the ellipsoid method, quite unrealiable, even though when it worked, it outperformed MINOS by a factor of 10 . A possible, and partial, explanation is that the introduction of constraints (no negative cycles, or $\lambda \geq 0$) leads to a vector field of subgradients which is highly non symmetrical.

## 5. Acknowledgements

This research benefited from the support of the FCAR (Quebec) and the NSERC (Canada).

## 6. Bibliography

[1] Agmon S. "The Relaxation Method for Linear Inequalities" Canadian Journal of Mathematics, pp. 382-392, 6 (1954)

[2] Bertsekas D.P., Gafni E.M. and Vastola K.S. "Validation of Algorithms for Optimal Routing of Flow in Networks", Proceedings 1978 IEEE Conference on Decision and Control, San Diego, January 1979.

[3] Ecker J.G. and Kupferschmid M. "A Computational Comparison of the Ellipsoid Method with SEveral Nonlinear Programming Algorithms", SIAM Journal on Control and Optimization, pp. 657-674, 23 (1985).

[4] Gafni E.M. and Bertsekas D.P. "Two-metric Projection Methods for Constrained Optimization", SIAM Journal of Control and Optimization, pp. 936-964, 22 (1984).

[5] Goffin J.L. "Convergence Results in a Class of Variable Metric Subgradient Methods" in Nonlinear Programming 4, Editors: O.L. Mangasarian, R.R. Meyer and S.M. Robinson, Academic Press (1981) pp. 283-326.

[6] Goffin J.L. "Convergence of a Cyclic Ellipsoid Algorithm for Systems of Linear Equalities", Mathematical Programming 22, pp. 239-260, (1982).

[7] Goffin J.L. "Convergence Rates of the Ellipsoid Method on General Convex Functions", Mathematics of Operations Research, pp. 235-250, 8 (1983).

[8] Goffin, J.L. "Variable Metric Relaxation Methods, Part I: A Conceptual Algorithm", SOL Technical Report 81-16, Department of Operations Research, Stanford University, Stanford, California, U.S.A.; 101 pp.(Aug. 1981).

[10] Held M. and Karp R.M. "The Taveling Salesman Problem and Minimum Spanning Trees: Part 2", Mathematical Programming, pp. 6-25, 1 (1971)

[11] Held M., Wolfe P. and Crowder H. "Validation of Subgradient Optimization", Mathematical Programming, pp. 62-88, 6 (1974).

[12] Hoffman A.J., Mannos D., Sokolowsky D. and Wiegman N. "Computational Experiments in Solving Linear Programs", SIAM Journal, pp. 17-33, 1 (1953).

[13] Khacian L.G., "A Polynomial Algorithm in Linear Programming", Z. Vycisl. Mat. Mat. Fiz., pp. 51-68, 20 (1980); translated in USSR Comp.Math.Math.Phys.,pp.53-73,20.

[14] Motzkin T. and Schoenberg I.J. "The Relazation Method for Linear Inequalities" Canadian Journal of Mathematics, pp. 393-404, 6 (1954).

[15] Shor N.Z. "Minimization Methods for Non-Differentiable Functions", Springer-Verlag,Berlin, (1985).

[16] Yudin D.B. and Nemirovskii A.S. "Estimation of the Informatinal Complexity of Mathematical Programming Problems", Econom. i Mat. Metody, pp.128-142, 12 (1976); translated in Matekon pp. 3-25, 13.

# ASYMPTOTIC STABILITY PROPERTIES OF SELF-ACCESSIBLE CONTROL SYSTEMS

Hernán R. Henríquez
Universidad de Santiago de Chile
Departamento de Matemática y Ciencia de la Computación
Casilla 5659 - Correo 2
Santiago - Chile

Abstract.

In this note we establish some relations between the concept of self-accessibility and the asymptotic behaviour of trajectories for linear time invariant systems.

## I. INTRODUCTION

Repeated processes occur naturally in a number of situations. The abstract formulation of this idea leads to the concept of self-accessibility. Here we will restrict us to consider time invariant systems:

$$\dot{x}(t) = A\,x(t) + B\,u(t) \tag{1}$$

where A and B are nxn and nxm matrices, respectively, and the admissible controls $u(\cdot)$ are continuous functions.

Let $t_1 > 0$. Following [ 1 ] we say that the system (1) is self-accessible on $[0,t_1]$ is for every initial state $x_o$ there exists an admissible control $u(\cdot)$, defined on $[0,t_1]$, such that the solution of (1) satisfies $x(0) = x(t_1) = x_o$.

It is clear that every controllable system is also self-accessible, but the converse assertion is false. Moreover, it is well know [4] that the controllability condition is equivalent to the arbitrary pole allocation.

The aim purpose of this note is to show that the self-accessible systems of type (1) verify a similar stabilizability condition and to related those systems with the controllables systems. Our development is based on the semigroup theory of operators, and not in canonical forms, for which the results obtained can be generalized to systems with states in infinite dimensional spaces.

## II. RESULTS.

Let be denoted by C to the controllability matrix of system (1), i.e.,

$$C = [\, B,\ AB,\ A^2B, \ldots\ldots,\ A^{n-1}\ B\,]$$

Also we reserve the italic letter $\mathcal{R}$ to denote the range space of linear operators.

### Theorem 1

The control system (1) is self-accessible on $[0,t]$, for every $t > 0$ if and only if $\mathcal{R}(A) \subseteq \mathcal{R}(C)$.

### Proof.

Considerer $t > 0$ and let be $G : C\,(0,t;\mathbb{R}^m) \longrightarrow \mathbb{R}^n$ the bounded linear operator defined by

$$G(u) = \int_0^t e^{A(t-s)}\ B\ u(s)\ ds.$$

Then the solution $x(t)$ to the equation (1), with initial condition $x_\circ$ is given by

$$x(t) = e^{At}\, x_\circ + G(u)$$

On the other hand, it is well known [2] that $\mathcal{R}(G) = \mathcal{R}(C)$. Therefore, the system (1) is self-accessible on $[0,t]$ if and only if $\mathcal{R}(I-e^{At}) \subseteq \mathcal{R}(C)$.

Now, we complete the proof remarking [ 3 ] that

$$e^{At} - I = A \int_0^t e^{As} \, ds$$

and the matrix

$$Q(t) = \int_0^t e^{As} \, ds$$

is invertible for $t$ sufficiently small.

The system (1) with the matrices

$$A = \begin{pmatrix} 0 & 1 \\ -1 & 0 \end{pmatrix} \quad \text{and} \quad B = \begin{pmatrix} 0 \\ 0 \end{pmatrix}$$

provides an elementary example of a system self-accessible on $[0,2\pi]$ but that's not self-accessible on $[0,t]$ for $0 < t < 2\pi$ . On the other hand, a system may be asymptotic stabilizable but not controllable. However these concepts are concerted with the following result.

<u>Theorem 2</u>

Let $t_1 > 0$. If the control system (1) is stabilizable and self-accessible on $[0,t_1]$ then it is also controllable.

<u>Proof</u>.

Let be $F$ a feedback matrix and consider the system

$$\dot{x}(t) = (A + BF) \; x(t) + Bu(t) \tag{2}$$

Since the solutions to the Cauchy problem:

$$\dot{x}(t) = A\,x(t) + BF\;x(t)$$
$$x(0) = x_0$$

are given by

$$x(t) = e^{At} \; x_0 + \int_0^t e^{A(t-s)} \; BF \; x(s) \; ds$$

it is clear that the system (1) is self-accessible (respectively, controllable) if and only if the system (2) is self-accessible (controllable). Therefore we may assume that the system (1) is asymptotically stable.
Thus, for every state $x \in \mathbb{R}^n$,

$$x - e^{At_1} \; x \in \mathcal{R}(C)$$

and since $\mathcal{R}(C)$ is a invariant subspace for $A$, hence for the exponential matrix $e^{At_1}$, we infer that

$$x - e^{nAt_1} \; x \in \mathcal{R}(C),$$

for every integer $n \geq 1$. Now, taking limit as $n \to \infty$, we obtain $x \in \mathcal{R}(C)$ and the proof is complete.

Finally, we will show that the self-accessesible control system s of type (1) have a asymptotic stability property, specifically:

Theorem 3

If the linear control system (1) is self-accessible on $[0,t]$ for every $t > 0$, then there exists a feedback matrix $F$ such that for every initial state $x_0$ the solution to

$$\dot{x} = (A + BF) \quad x(t) \qquad (3)$$
$$x(0) = x_0$$

is convergent as $t \to \infty$.

Proof.

Let $E$ be a suplementary subspace of $\mathcal{R}(C)$, i.e,

$$\mathbb{R}^n = \mathcal{R}(C) \oplus E$$

Next we denote by $x_1$ and $x_2$ the components of the vector $x$ in $\mathcal{R}(C)$ and E, respectively. Now, by the theorem 1 we may define the linear operators.

$$A_1 = A|_{\mathcal{R}(C)} \quad : \quad \mathcal{R}(C) \longrightarrow \mathcal{R}(C)$$

and

$$A_2 = A|_E : E \longrightarrow \mathcal{R}(C),$$

and since $\mathcal{R}(B) \subseteq \mathcal{R}(C)$, the control system (1) may be decomposed in two subsystems:

$$\dot{x}_1(t) = A_1 x_1(t) + A_2 x_2(t) + B u(t) \tag{5}$$

$$\dot{x}_2(t) = 0, \tag{6}$$

hence $x_2(t) = x_2(0)$ and the system (5) is reduced to

$$\dot{x}_1(t) = A_1 x_1(t) + B u(t) + A_2 x_2(0) \tag{7}$$

on the other hand, it is clear that

$$\mathcal{R}[B, A_1B, \ldots\ldots, A_1^{n-1} B] = \mathcal{R}(C),$$

so that [ 2 ] the linear control system

$$\dot{x}_1(t) = A_1 x_1(x) + B u(t) \tag{8}$$

is controllable. Now, from the well known Wonham's stabilizability result [ 4 ] we obtain that there exists a linear operator $F_1 : \mathcal{R}(C) \longrightarrow \mathbb{R}^m$ such that the eigenvalues of the linear operator $A_1 + BF_1$ have negative real part.

Therefore, defining $u(t) = F_1 x_1(t)$ the solutions to the equation (7) verify ( [ 3 ] ),

$$\lim_{t \to \infty} x_1(t) = -(A_1 + BF_1)^{-1} A_2 x_2(0)$$

Therefore, it is suficient to define the operator $F : \mathbb{R}^m \longrightarrow \mathbb{R}^n$ by $Fx = Fx_1$

to obtain the desired result.

## REFERENCES

[1] Baccioti, A., "Auto-acessibilité par familles symetriques de champs de vecteurs", Ricerche di automatica, 7,189-197, 1976.

[2] Lee, E.B.; Markus, L., Foundations of optimal control theory, John Wiley & Sons, New York, 1967.

[3] Pazy, A., Semi-groups of linear operators and applications to partial differential equations, Lecture Notes N° 10, University of Maryland, 1974.

[4] Wonham, W.M., "One pole assignment in multi-input controllable linear systems", I.E.E.E. Trans. Autom. Control, Vol. AC-12, 6, 660-665, 1967.

# MATHEMATICAL MODELS FOR PINE TIMBER PRODUCTION PLANNING

Ramiro Morales, Facultad Ciencias Forestales, U. de Chile
Andrés Weintraub, Depto. Ingeniería Industrial, U. de Chile

## 1. INTRODUCTION

Long-term planning is a basic requirement of forestry. Usually, it implies planning for a time interval of one rotation length or more, that is, more than 24 years for fast growing species.

The significance of long-term planning depends largely upon the size of the forest area under consideration. Thus, if the decision-makers are small forest owners, their time horizon is typically short and a great deal of uncertainty is associated with their future timber production activities. From their point of view, long-term planning is of low or doubtful value. However, at a regional or national level, or at the level of industrial enterprises, long-term planning is imperative and profitable, considering the scale of investments involved.

In Chile, forest industrial enterprises are based, mainly, on a timber resource composed of radiata pine plantations. Each firm owns a total forest area that ranges from 25.000 to 150.000 hectares. These firms are faced with forests composed of different planting ages, locations and silvicultural states; with a large number of potential treatments for each stand, which vary with the type, intensity and timing of forest practices; and with a complex set of timber requirements. Under these conditions, long-term planning needs effective and efficient tools for evaluating different policy alternatives.

The use of mathematical models in forestry decision-making is being extensively adopted by chilean forest industrial enterprises that manage radiata pine plantations. The extent to which mathematical models are being adopted, and the type of models used, are determined by the quality of the available data. These data comprise those that describe the present condition of the stock of plantations, growth data for predicting the evolution of the stands after implementing silvicultural practices, and cost and revenue estimates.

This paper presents chilean experience with mathematical models for radiata pine timber production planning in industrial enterprises. A stand simulation model is presented in section 2. This model is intended to predict growth and timber yields resulting from thinning practices, generating a set of data that indicates the expected states of a radiata pine stand over time. The results may be utilized as input data for higher-level planning models, such as simulation of timber production of entire plantations (section 3), or, optimization models for plantation management (section 4).

## 2. GROWTH SIMULATION MODEL.

The simulation of growth and yield of radiata pine stands has been focused to answer two interdependent questions: the optimal initial planting density and the optimal frecuency and intensity of thinning practices. Both questions are related to the form of distributing the potential wood production of the stand among its individual trees.

There is an abundant literature on growth prediction under different silvicultural practices. [1, 4, 5, 8, 14, 15]. A detailed survey and discussion is presented in [12].

The models involve different functional equations, according to silvicultural characteristics and criteria considered. For pine plantations in Chile, a particular set of equations was developped, (RADIATA) for different geographical regions and programmed as a computer package available to users.

RADIATA is a simulation model designed to predict the growth and timber yields of radiata pine stands under a wide range of spacing options. A stands is considered as a homogeneous area for management. In pine plantations, location, age of trees, site quality and density conditions are the main elements to define a stand. It generates standing timber production data for each stand and thinning alternatives over the planning horizon so that the best management option can be chosen in the optimization phase.

The variables which characterize the state of the stand at a point t in simulated time are:

$E_t$ : stand age at t (years).

$HDOM_t$ : arithmetic mean height of the 150 largest trees (by diameter) per hectare at t (m).

$N_t$ : number of trees per hectare at t.

$G_t$ : stand basal area at t ($m^2$/ha).

$D_t$ : diameter (at 1.3 m height) corresponding to the tree of mean basal area at t (cm).

$VX_t$ : standing timber volume at t ($m^3$/ha) considering an upper stem diameter of X cm (X = 10; 15; 20 and 25 cm).

The initial state of the stand is specified by $E_0$, $HDOM_0$, $N_0$ and $G_0$. The initial values assumed by these state variables are taken from field measurements.

If the user wants to simulate the effect of thinning practices, the model is able of predicting timber output and future stand growth resulting from these partial cuttings. Stand age(s) at the time of thinning and cutting intensity are input data of the model. Thinning intensity can be specified either in number of trees/ha or in terms of basal area/ha to be left after each partial cutting.

Stand growth is described by a set of transition functions (difference equations) which project $HDOM_t$, $N_t$, and $G_t$ in time. These functional relationships were derived using regression analysis on actual data coming from permanent sampling plots distributed on three geographical areas. The data was provided by the Chilean Forest Institute. It was derived from sample plots showing regular spacing, no damage from biotic or abiotic agents and negligible brush competition. The data base is composed of 166 unthinned plots and 94 thinned plots established on a wide range of age, number of trees and site quality, in 1962.

Dominant height ($HDOM_t$) is estimated by $HDOM_{t+\theta} = HDOM_t \exp\{\alpha(1/E_t - 1/E_{t+\theta})\}$ where $\alpha$ varies between 12.4 and 14.8 according to the geographical area being considered, and $\theta$ is the length of the time interval.

To predict natural tree-mortality and basal area growth the model uses functions that describe observed patterns with high fidelity. Different equations are used depending on whether the stand is thinned or not.

The state variable $VX_t$ is calculated at the end of each time interval as a function of $HDOM_t$, $N_t$ and $G_t$. Another set of equations is used to express thinning intensity in terms of timber volume harvested and to estimate the state of the stand after partial cuttings.

RADIATA has been implemented for three geographical regions of distribution of radiata pine in Chile, namely, Concepción-Arauco, Arenales y Maule. The following functional relationships were derived for the Maule region:

Unthinned stands:

$$N_{t+3} = \begin{cases} N_t & \text{if } N_t \leq 5596.4 - 1952.1\, L_n(HDOM_t) \\ 273.6650 + 0.9511\, N_t - 95.4599\, L_n\, (HDOM_t), & \text{otherwise} \end{cases}$$
$$(R^2 : 0.984 \quad S.E.: 34.0)$$

$$G_{t+3} = 26.4284 + 1.0359\, G_t - 6.4808\, L_n\, (HDOM_t) - 0.4760\, G_t\, (N_t - N_{t+3})/N_t$$
$$(R^2 : 0.995 \quad S.E.: 1.57)$$

$$V10_t = -0.0160\, N_t + 2.3386\, G_t + 0.2580\, G_t\, HDOM_t$$
$$(R^2 : 0.998 \quad S.E.: 7.4)$$

Thinned stands:

$$N_{t+2} = -3.2670 + 0.9973\, N_t$$
$$(R^2 = 0.996 \quad S.E.: 11.1)$$

$$G_{t+2} = \begin{cases} 6.2618 + 1.0571\, G_t - 46.3077\, G_t/N_t \\ \text{(2 years after thinning)} \quad (R^2 : 0.996 \quad \text{S.E.: } 0.60) \\ \\ -7.6993 + 0.9492\, G_t + 2.3889\, L_N(N_t) - 0.9361\, G_t(N_t - N_{t+2})/N_t \\ \text{(2 years after thinning)} \quad (R^2 : 0.984 \quad \text{S.E.: } 1.56) \end{cases}$$

$$V10_t = -7.1855 + 0.2308\, G_t\, HDOM_t + 0.10171\, G_t\, D_t \quad (R^2 : 0.996 \quad \text{S.E.: } 8.89)$$

Thinning

$$V10R_t = V10B_t\,(GR_t/GB_t)^{1.0758} \quad (R^2 : 0.981 \quad \text{S.E.: } 0.10)$$

$$V10R_t = 0.3527\,(V10B_t\, NR_t/NB_t)^{1.101} \quad (R^2 : 0.843 \quad \text{S.E.: } 0.13)$$

R: removed B: before thinning

For thinned and unthinned stands:

$$VX_t/V10_t = F(HDOM_t, V10_t, D_t)$$

The correlation coefficients standard errors, F and t values of the above regression equations show their excellent adjustment to the data base.

Some difficulties arise in trying to compare the results generated by RADIATA with real data. No work describing the growth and yield of Pinus radiata has been yet published to the extent of making possible rigorous comparisons with other findings. Simulation with real data input, on the other hand, requires information from repeated measurements of permanent growth plots which, with the exception of the data used to construct the model, is not available.

What can be said is that RADIATA replicates very closely the original data used to estimate the parameters of the prediction equations, and that the model is consistent with suggested theory of plantation growth. The capability of the equations for explaining the variations in growth and yield decreases with the quality of the forest site, though it remains high enough to provide reasonable predicting accuracy. Correlation analysis, including plots of residuals, failed to show any significant anomaly.

RADIATA is being used in several ways. It generates the technical coefficients for timber yields under different management alternatives to input into LP models. It is also used in global simulation (section 3) and, as information to determine optimal management policies at the stand level, as shown in [13].

3. SIMULATION OF TIMBER PRODUCTION FOR ENTIRE PLANTATIONS.

Several models for simulating timber production of entire plantations are being used by forest industrial enterprises in Chile.

Practically all of them are designed to determine the allowable cut that can be sustained in a given time interval, usually one rotation length.

The models predict the growth and timber yields, under different treatments, for a set of stands of similar characteristics using local yield tables or stands simulators, such as RADIATA. The schedules and intensities of silvicultural practices and the cutting priority are fixed for the run.

The main application of this type of models are in long-term planning to estimate the balance between timber availability and the quantity of wood demanded by present and future mills. The results are aggregated and do not contain economic aspects.

The advantages of this type of models in relation to optimization techniques (section 4) are that the former are faster to run on the computer and therefore cheaper; they can be run in microcomputers in inter-active mode; and they are more accepted by decision-makers, most of whom do not have experience about optimization methods.

Simulation models for answering forest-wide questions do not seek to optimize cutting plans but to search for an approximate allowable cut. To determine an optimal cutting plan would require a large number of trials and finding a solution may be very difficult.

Simulation models can often identify feasible solutions for cutting plans and, in this sense, their results may be used as information to feed optimization models.

4. OPTIMIZATION MODELS IN PINE PLANTATION MANAGEMENT.

It is possible to search for optimal management at stand level by the use of simulation, by testing through a program like RADIATA a series of alternative treatments and making an economic evaluation of the resulting yields. However, there will normally exist an interaction among the different stands, in terms of total timber yield requirements, financial aspects, access problems, etc. These interactions make it imperative to consider global management policies for plantations as a whole in planning. Simulation presents severe shortcomings in determining optimal or near optimal policies for such global problems. Hence, the need to introduce optimization tools.

In [2], a survey of modeling approaches to problem solving in different forest sectors is presented. The literature ([10]) reports on two basic timber production optimization models. Model I ([11]) defines management activities (thinnings or partial cuts, clearcuts, etc.) associated to specific land areas. An activity in Model I is specified by the volume/ha obtained in each period through the horizon for each management activity defined.

Model II defines activities associated to the period of planting and clearcutting. In this form, timber classes are regrouped as cuttings occur and the spacial definition of classes is lost. Model I is more intuitive to use and easier to implement, but defining an adequate number of management alternatives for each timber class may lead to a very large number of variables. Model II requires a smaller number of variables (one per combination of year of planting and clearcutting). If however, intermediate interventions are defined, and site quality and location considered, the number of variables also becomes large.

The present trend, not yet fully developped is towards a combination of both concepts [7, 9]. At early periods, when a higher degree of resolution is required, a Model I type of approach is used. For latter periods, when higher aggregation and approximation are allowed, areas are switched to a Model II type, defining just a few alternative variables for each combination of period of planting and clearcut to allow for alternative possibilities of intermediate interventions, site locations, etc.

Most applications so far follow the Model I approach though. Models designed specifically for pine plantations in private industry have been published (Australia [6], Chile [3]) including the case of vertically integrated industries.

In Model I, activities or management alternatives are expressed by the timber yields per acre obtained in different periods due to interventions. In this form, total production for a period t is defined by:

$$H^t = \sum_{ij} a_{ij}^t x_{ij}, \text{ where}$$

$x_{ij}$ = number of ha of stand i managed with alternative j;

$a_{ij}^t$ = yield ($m^3$/ha) obtained in period t with that alternative.

Note that $a_{ij}^t$ will be zero for most periods. Analogously $c_{ij}^t$, $d_{ij}^t$ can be the cost and revenues per acre associated to alternative (ij) in period t.

The above defined are basic variables and parameters to define a land holding planning linear programming model. Note that in any solution, $\sum_j x_{ij}$ equals the area of stand i. Due to the characteristics of the basis in LP solutions, normally for each stand only one or two activities will be at positive level. Periods can be defined yearly at the beginning and lengthened towards the hori̲zon. The planning horizon is taken usually to include about two cutting cycles. An objective function might be to maximize net present revenues, with constraints on minimum production per period. Other possible constraints [11] might require non-decreasing yields, $H^t \geq H^{t-1}$ for all t.

In [3], 25 experimental stands were considered in a model for a land holding firm. About 10-12 management alternatives were defined for each stand. The growth simulation tests carried out through RADIATA allowed to analyze which were potencially attractive alternatives to include in the model. Programs we̲re defined to maximize net present revenues subject to a minimum level of sales per period. In previous reported experience, not much emphasis has been given to post-optimal analysis. In this case, sets of problems were run to test in particular:

a) The impact of variations in future timber market prices. The results are of interest not only in terms of revenues obtained, but also to analize the robustness of the solutions obtained. If there is an error in the estimation of future prices, what does this imply in terms of the management decisions for the first few periods, which are the ones that are actually implemented? The results indicated that in most cases, there was no drastic difference in stand management and thus an error in estimation of future prices would lead to a relatively low degradation of the objective function due to misguided inicial decisions.
b) Boundary conditions. Typically, due to institutional objectives, which go beyond maximizing net present worth, restrictions are imposed in some forms at the horizon, ir order to insure continuity of the firm. One simple way of doing this is through requiring a certain distribution of stand ages at the horizon. In the problems tested, these constraints did not significantly reduce the objective value at the planning horizon (50 years), given the effect of the discount rate and the rationality in the selection of possible management alternatives for the L.P.

Comparisons were made using efficient stand management alternatives in a straightforward form in a simulation process. The interaction among stands was given by requirements of minimal sales per period. The results obtained through simulation gave substantially lower revenues than using an optimization procedure. This can be explained through the difficulties in allocating in an efficient way intermediate interventions, and total production per year with simple simulation approaches. More sophisticated simulation rules reduce these problems, though not completely, but require a significantly larger design and computational effort.

In the case of vertically integrated firms, that is firms which complement land holdings with industrial facilities such as pulp and sawmills, the decisions involve timber management in the stands and production at the industrial facilities, e.g. how much timber is sent to the pulp plant, sawmill or sold as logs. There is usually also the option of purchase of timber to outside suppliers.

This problem was modelled in a real industrial decision making context, [3]. Industrial facilities were considered as black boxes, which transform timber into products at a certain cost. To estimate net revenues, these transformation costs plus costs due to transportation to market places, management at stands and hauling to plants are deducted from sales.

Typical constraints are:

a) Institutional: financial requirements, minimal sales to certain markets, timber cutting policies (e.g. non-declining yields).
b) Technical: plant capacities, production characteristics (e.g. residues of sawed timber can be used as input for the pulp plant).
c) Market: sales limitations, bounds on timber purchases from outside suppliers.

Again, post-optimal runs were made to investigate the behaviour of solutions in terms of:

a) Increase of plant capacities. This is of interest to evaluate the profitability of expansion investments.
b) Purchase of timber lands. The increased value of the solution including the new areas allows to evaluate the possible acquisition.
c) Minimum production levels imposed for certain products (e.g. imposing a minimum level production of sawed timber led to substantial losses in revenues). This analysis is typical when considering long term delivery contracts.
d) Future price variations. This case is more complex than for pure land holdings, as the relative prices among products will vary. Again, solution were found to be quite robust.

Models currently published or in use are essentially suited for tactical planning. Strategic decisions such as land acquisition or plant building can only be approached by changing parameters in these models, one at a time as indicated in the post-optimal analysis above. When there is a set of possible such decisions, a mixed integer program should be used where integer 0-1 variables represent discrete investment alternatives.

A two state hierarchical planning approach has been proposed [16]. At strategic level, a mixed integer LP program concentrates on investment decisions. Stands, management alternatives and time periods are aggregated in order to reduce the size of the problem. Operational and tactical decisions (forest and plant operations) are evaluated in a lower level model such as in [3]. Global production resulting from the strategic model become targets for the tactical model in order to preserve consistency in decisions. This is not trivial, as the aggregation process may induce to errors. LP aggregation procedures, first proposed by Zipkin [17] has proved a helpful tool in this process. Strategic planning models have not yet been implemented in actual decision making, but present a coming challenge both in research and application.

In order to allow an easy use of the described models they must be inserted within a computer package. For example, in the case of a private firm, a data base corresponding to the firm holding was developed [3]. In this situation, the user simply inserts basic parameters to describe the problem he wishes to run. The data base is updated as needed.

A more flexible tool was developed in [13], where the user generates in an interactive form the data to be used. The main elements of the package are:

a) Management alternatives generator. It is based on RADIATA. The user specifies which set of growth prediction equations is best suited to his stands.
b) Matrix generator, which incorporates needed additional data (costs, production relations, etc.) and sets up the problem in LP format.
c) Report writer, which presents the results in a legible form, as well as allows analysis through tables and graphs.

## 5. CONCLUSION.

The use of mathematical models both for growth prediction and as support for timber harvest management has been used by forest enterprises in Chile. The advantageous use of such models in this country is clearly feasible, as the recent initial experiences described have shown. Our experience has been that the main conditions for establishing the use of such models are:

a) The development of data on growth over a number of years under different practices, possibly using experimental stands, in order to have acceptable growth simulation models. These are basic to any further accurate quantification of the problem.

b) The existence of analysts capable of implementing these models and management aware of its benefits and limitations.

REFERENCES.

[1] Alder, D. 1978. PYMOD: A forecasting model for conifer plantations in the Tropical Highlands of Eastern Africa. In: Fries, J.; H.E. Burkhart and T.A. Max (editors). FWS-1-78 School of Forestry and Wildlife Resources. Virginia Polytechnic Institute and State University.

[2] Eare, BB, Briggs, G., Roise, J.P., Schrender, G.F. A survey of systems analysis models in forestry and the forest products industries. European Journal of Operational Research. Volume 18, N° 1, pp. 1-18. 1984. North-Holland-Amsterdam.

[3] Barros, O. and A. Weintraub. Planning in forest enterprises. Operations Research. Vol. 30, N° 6, pp. 1168-1182. (1982).

[4] Beekhuis, J. 1966. Prediction of yield and increment in Pinus radiata stands in New Zealand. New Zealand For. Ser. Wellington, Tech. Pap. N° 49.

[5] Clutter, J.L. y D.M. Belcher. 1978. Yield of site-prepared slash pine plantations in the lower costal plain of Georgia and Florida. In: Growth models for long term forecasting timber yields. IUFRO, Working Group S.4.01. Fries et al. (editors).

[6] Dargavel, J.B. A model for planning the development of industrial plantations. Australian Forestry. Vol. 4, N° 2 (1978).

[7] Franetovic, P. A model for managing forest plantations. Engineering thesis. Dept. of Industries, U. de Chile (1980).

[8] Hann, D.W. and Brodie, J.D. Even-aged management: basic management questions and availability of techniques for answering them. USDA Forest Service, GRT INT-83, Ogden, UT. 1980.

[9] Johnson, K.N. at al. 'FORPLAN' User's Manual.US Forest Service, 1979.

[10] Johnson, K.N. and H.L. Scheurman. 1977. Techniques for prescribing optimal timber harvest and investment under different objectives: Discussion and synthesis. Forest Science Monograph 18.

[11] Navon, D.I. 1971. Timber RAM, a long range planning method for commercial timber land under multiple-use management. USDA For. Ser. Res. Paper

PSW 70, Berkeley, California.

[12] Morales, R., Weintraub, A., Peters, R., García, J. Modelos de Simulación y Manejo para plantaciones forestales. Documento N° 30, Proyecto CONAF/PNUD/FAO. CHI/76/003. Santiago de Chile, 1979.

[13] Morales, R., Weintraub, A., Olivares, B., Peters, R. Modelo para el Manejo de Pino Insigne. Documento N° 36. Proyecto CONAF/PNUD/FAO. CHI/76/003. Santiago de Chile, 1981.

[14] Stage, A.R. 1973. Prognosis model for stand development. USDA Forest Service Report INT-137.

[15] Turner, B.J.; R.W. Bednarz and J.B. Dargavel. 1977. A model to generate stand strategies for intensively managed radiata pine plantations. Australian Forestry, 40(4).

[16] Weintraub, A., Guitart, S. and V. Kohn. 'Investment Planning in Forest Industries: A Hierarchical Approach.' To appear in European Journal of Operations Research.

[17] Zipkin, P. Bounds on the Effect of Aggregating Variables in Linear Programs. Operations Research. Vol. N° 2, N° 3. pp. 403-418 (1980).

# A CONTROL PROBLEM WITH A NON-COERCIVE COST FUNCTIONAL

P.R. Oliveira
Instituto de Matemática and COPPE
Universidade Federal do Rio de Janeiro
21.944 - Rio de Janeiro - Brazil

## 1. Introduction

In this paper we analyse a control problem that has a $L^1$-type cost functional. This functional is non coercive in Hilbert spaces, and we do not have prior existence and unicity solutions. However there is convexity, and therefore the duality methods of Convex Analysis can be exploited.

The controlled system has a parabolic equation with initial and Dirichlet conditions as given in a paper by Bensoussam, Nissen, Tapiero [1], where the aging phenomenon is modelled. In their paper a typical quadratic cost function is associated with the equation, and Control Theory is used to get the optimal replacement by Riccati with a linear feedback. Now, as we have a constrained control variable, this is no longer possible.

Convex Analysis provides optimality conditions, but it is necessary to define and analyse an inverse problem, here an integral equations , to prove the existence of solution, and obtain an "almost explicit" form.

## 2. The problem

Consider, for $T > 0$, $\Omega = (0,1)$, $Q = \Omega \times (0,T)$, the parabolic equation

$$y_t + \alpha\, y_x - \varepsilon y_{xx} = 0 , \quad (x,t) \in Q, \tag{1}$$

$\alpha > 0$ and $\varepsilon > 0$ being constant coefficients, with the initial and bondary conditions:

$$y(x,o) = y_o(x),\ x \in \Omega \tag{2}$$

$$y(0,t) = v(t), \quad y(1,t) = 0, \quad t \in (0,T) \tag{3}$$

If the initial condition yo is in $L^2(\Omega)$ and $v$ is in $L^2(0,T)$ we know that (1), (2), (3) is well posed in a certain weak form, and the unique solution y is in $H^{1/2,1/4}(Q)$. See [2].

Associated with this boundary - condition problem consider the cost functional

$$J(v,y(v)) = \int_0^T cv(t)dt + \int_0^T \left| \int_0^1 y(x,t,v)dx - Z(t) \right| dt, \tag{4}$$

where, for a given $v$, $y$ is the solution of (1), (2), (3), $Z$ is a non negative real function on $[0,T]$ (the demand function), and c is a real positive number. $v$ is the control variable, the range of which is

$$L^2_+(0,T) = \{v,\ v \in L^2(0,T),\ v \geq 0 \text{ a.e.on } (0,T)\} \tag{5}$$

Then the control problem is

$$(P) \inf J \text{ over } L^2_+(0,T) \tag{6}$$

3. <u>The dual problem</u>

The generality of (6) is not lost if the initial condition (6) is made zero:

$$Y_0 \equiv 0 \tag{7}$$

(formally, it is sufficient to change $Z$ in $J$ by $Z(t) - \int_0^1 y^0(x,t)dx$, where $Y^0$ is the solution of (1), (2), (3) with $v \equiv 0$). Now define

$$F(v) = \begin{cases} \int_0^T cv, & \text{if } v \in L^2_+(0,T) \\ +\infty, & \text{if not} \end{cases} \tag{8}$$

$$G(Av) = \int_0^T |Av - Z|, \quad v \in L^2(0,T), \tag{9}$$

where the linear operator $A : L^2(0,T) \longrightarrow L^2(0,T)$ is given by

$$Av = \int_0^1 y(x,t,v)dx, \tag{10}$$

y being the solution of the parabolic system (with $y_o \equiv 0$). Then we have (6) written as

$$(P) \inf J(v,Av) = F(v) + G(Av), \quad v \in L^2(0,T) \tag{11}$$

As it is known from duality theory (see [3] and [4]), the dual problem can be defined by

$$(P^*) \sup \{ - F^*(q) - G^*(-p)\}, \quad p \in L^2(0,T), \tag{12}$$

where $F^*$ and $G^*$ are, respectively, the polar functions of F and G, p and $q = A^*p$ are dual variables, $A^*$ is the adjoint operator of A:

$$F^*,\ G^*: L^2(0,T) \longrightarrow \bar{R}\ , \quad A^*: L^2(0,T) \longrightarrow L^2(0,T) \tag{13}$$

Using the formula $F^*(9) = \sup \{(v,q)_{L^2(0,T)} - F(v)\}$ a straightforward calculation provides (see [3], [5] and [6]):

$$F^*(q) = \begin{cases} 0 \ , & \text{if } q \leq c \\ +\infty \ , & \text{if not,} \end{cases} \tag{14}$$

and, similarly,

$$G^*(-p) = \begin{cases} -\int_o^T pZ \ , & \text{if } -1 \leq p \leq 1 \\ +\infty \ , & \text{if not} \end{cases} \tag{15}$$

Finally we get

$$(P^*) \sup \int_o^T pZ \ , \quad p \in K, \tag{16}$$

where

$$K = \{p,\ p \in L^2(0,T),\ -1 \leq p \leq 1,\ q(p) \leq c,\ q(p) = A^*p\}.$$

*The adjoint operator $A^*$*

Define the adjoint system of (1), (2), (3) as

$$\left|\begin{array}{ll} -r_t - \alpha r_x - \varepsilon r_{xx} = p & ,\ (x,t)\varepsilon Q \\ r(x,T) = 0, \quad x\,\varepsilon\,\Omega & \\ r(0,t) = r(1,t) = 0, t\ \varepsilon\ (0,T) & \end{array}\right. \tag{17}$$

Observe from (10) that by multiplying Av and p in $L^2(0,T)$ we have

$$(p,\ Av) = \int_Q py \tag{18}$$

If we use this relation and (17), we get, integrating by parts

$$(p,Av) = \int_Q (-r_t - \alpha r_x - \varepsilon r_{xx})\ y = \int_0^T \varepsilon r_x(0,t;p)v(t)dt = (A^*p,v) \tag{19}$$

Now suppose p is sufficiently regular, e.g., $p\varepsilon\ H^{3/8}(0,T)$, a trace theorem (see [2]) applied to (17) assures the continuity of $r_x(0,t;p)$ on $[0,T]$. Later results will confirm that this is a correct assumption, the dual solution $\bar{p}$ being a continuous function. Thus we can write

$$A^*p(t) = \varepsilon\ r_x(0,t;p) \quad ,\ t\varepsilon[0,T] \tag{20}$$

which, with (16) and (17), defines completely the dual problem.

## 4. Some preliminary results

The first two results below are immediate applications of the general theory of convex analysis (see [4]). Propositions 3 to 5 give some insight in our specific problem, and are of easy demonstration. For the details, see [7].

*Proposition 1*: $P^*$ has at least one solution $\bar{p}$ and $\inf P = \max P^*$.

*Proposition 2*: The following conditions are equivalent:

(i) $\bar{u}$ is solution of P, $\bar{p}$ is solution of $P^*$ and $\min P = \max P^*$

(ii) $\bar{u}\ \varepsilon\ L^2_+\ (0,T)$ and $\bar{p}\ \varepsilon\ K$ verify the *transversality relations*:

$$(c - A^*\bar{p})\bar{u} = 0 \quad \text{a.e} \tag{21}$$

$|A\bar{u} - Z| + \bar{p}\,(A\bar{u} - Z) = 0$ a.e. (22)

*A simple case*

We will discard a very simple situation. Put, for the constant function 1 on $[0,T]$:

$$\gamma(.) = A^*1(.) \tag{23}$$

*Corollary 1*: Let $\gamma \leq c$. Then

$$\bar{u} = 0, \quad \bar{p} = 1 \tag{24}$$

are solutions of P and $P^*$, respectively. Moreover $\bar{u}$ is unique if $\gamma < c$ and $\bar{p}$ is unique if $Z > 0$.

Proof: Use (16), (17) and the transversality relations.

*An "almost explicit" solution*

For convenience we consider A a linear positive operator (it maps the positive cone of $L^2(0,T)$ into the positive cone of $L^2(0,T)$). We suppose also that the demand Z is to be reached, i.e., some $f \in L^2_+(0,T)$ must exists such that $Af = Z$. With these we have:

*Proposition 3*: If A is positive and $Z \in A(L^2_+(0,T))$, with $Z = Af$ for some $f \in L^2_+(0,T)$, then $P^*$ can be rewritten as

$$(P^*) \max \int_0^T qf, \quad q\in K_q, \quad K_q=\{q,\ q\in L^2(0,T),\ q\leq c \text{ a.e. } -1\leq p\leq 1,\ q=A^*p\} \tag{25}$$

This other form of $P^*$ is evident from $(p,Z) = (p,Af) = (q,f)$.

*Remark:* $A^*$ is positive if A is positive; thus, for all $q \in Kq$, it is evident that

$$q \leq \inf\{c, \gamma\}. \tag{26}$$

*Proposition 4:* The same assumptions as in Prop. 3, and Kq has a maximal function $\bar{q}$ (for the standard order relation), then $\bar{q}$ is a solution of $P^*$, given by (25).

*Proposition 5:* The same assumptions as in Prop. 4; moreover $\gamma$ is a continuous non-increasing function and there exists $\bar{t} \in (0,T)$ such that $\gamma(\bar{t}) = c$, then

$$\bar{q} = cX(0,\bar{t}) + \gamma X(\bar{t},T) \tag{27}$$

$$\bar{u} = fX(0,\bar{t}) \tag{28}$$

are solutions of $P^*$ and $P$, respectively.

The solution $\bar{q}$ is a direct consequence of the remark, the preceding proposition and the hypothesis; then $\bar{u}$, given by (28), can be verified to satisfy the transversality relations in Prop. 2.

## 5. The main results

The aim of this paragraph is to demonstrate that the foregoing propositions 3 to 5 are applicable to $P^*$ and $P$, so that (27) and (28) are their respective solutions.

*Lemma 1.* A, defined by (10), is a positive operator.

The proof is a straightforward use of the minimum principle for uniformly parabolic operators.

*Lemma 2.* $\gamma$ is a continuous non increasing function on $[0,T]$, and $\gamma(T)=0$. Consequently either $\gamma(t)<c$ on $[0,T]$, or there exists $\bar{t}$ such that $\gamma(\bar{t}) = c$.

Proof: The continuity is assured, as was already said, by a trace theorem.

Now apply the minimum principle to (17) with $p \equiv 1$; we have $a(x) = r_t\big|_{t=T} \leq 0$, the regularity of the solution justifying this trace. Put $Z = r_t$ and derive (17) in $t$ (for $p \equiv 1$), term by term; this is possible, the argument is, as always, the regularity of $r$:

$$\left|\begin{array}{ll} -Z_t - \alpha Z_x - \varepsilon Z_{xx} = 0 , & (x,t) \in \Omega \times (0,T) \\ Z(x,T) = a(x) , & x \in \Omega \\ Z(0,t) = Z(1,t) = 0 , & t \in (0,T) \end{array}\right. \tag{29}$$

The maximum principle applied to (29) gives $Z_x\big|_{x=0}\leq 0$, which is the same as $r_{xt}\big|_{x=0}\leq 0$ ; therefore $\gamma = r_x\big|_{x=0}$ is a non increasing function. We have also $\gamma(T) = 0$ since $r_x(x,T) = 0$ for all $x$ by the final condition $r(x,T) = 0$ in (17).

The second part of the lemma is obvious.

In order to apply propositions 3 to 5, it remains to show that $K_q$ has a maximal function $\bar{q} = \varepsilon r_x(0,t;\bar{p})$, which will be given by (27). With this purpose we define an inverse problem as follows: for an arbitrary $\bar{t}\varepsilon(0,T)$, consider the auxiliary system

$$\left|\begin{array}{ll} -y_t - \alpha y_x - \varepsilon y_{xx} = 0, & (x,t)\varepsilon\, \Omega x(0,\bar{t}) \\ y(x,\bar{t}) = 0, & x\,\varepsilon\,\Omega \\ y_x(0,t) = 0, \quad y(1,t) = p(t) \quad , & t\varepsilon(0,\bar{t}) \end{array}\right. \tag{30}$$

We note $L: H^{\frac{3}{4}}(0,\bar{t}) \longrightarrow H^{\frac{3}{4}}(0,\bar{t})$ the linear operator

$$L: p \longrightarrow y(0,.) \tag{31}$$

Furthermore for $p\equiv 1$, let $r(x,\bar{t})$ be the solution $r$ at $t = \bar{t}$ of the adjoint system (17). Then put

$$h(t) = \zeta(0,t), \tag{32}$$

where $\zeta(x,t)$ is the solution of

$$\left|\begin{array}{ll} -\zeta_t - \alpha\,\zeta_x - \varepsilon\,\zeta_{xx} = 0\,, & (x,t)\varepsilon\,\Omega\, x\,(0,\bar{t}) \\ \zeta(x,\bar{t}) = -\alpha r_x(x,\bar{t}) - \varepsilon r_{xx}(x,\bar{t}) - 1\,, & x\,\varepsilon\,\Omega \\ \zeta_x(0,t) = c/\varepsilon\,, \quad \zeta(1,t) = 0\,, & t\,\varepsilon\,(0,\bar{t}) \end{array}\right. \tag{33}$$

Now consider the affine operator $F$

$$F = h + Lp : \quad H^{\frac{3}{4}}(0,\bar{t}) \longrightarrow H^{\frac{3}{4}}(0,\bar{t}) \tag{34}$$

We have the following

*Lemma 3:* If $F$ has a fixed point $\bar{p}$, such that $-1 \leq \bar{p} \leq 1$ and

$$\bar{p} = h + L\bar{p} \tag{35}$$

then $K_q$ has a maximal function $\bar{q} = q(.,\bar{p})$, with $\bar{p} \in K$.

Proof: Observe that (17) is a time regressive system, thus, given $\bar{t} \in (0,T)$, if $\bar{p}$ is such that

$$\bar{p}/(\bar{t},T) \equiv 1, \tag{36}$$

this gives, by (17)

$$\bar{q}(t) = \varepsilon r_x(0,t;1) = \gamma(t), \qquad t \in (\bar{t},T), \tag{37}$$

which agree with the maximality constraint (26) and the proposed solution (27).

It remains to analyse, in the interval $(0,\bar{t})$, the inverse operator that maps $p$ into $q(p) = \varepsilon r_x(0,t;p)$, and to show the existence of $\bar{p}$ such that $q(\bar{p}) = c$. We will have once more (26) and (27).

The idea is to transform (17) by substituting the boundary condition $r(0,t) = 0$ into $r_x(0,t) = c/\varepsilon$, and by using the variable transformations

$$\pi(t) = -\int_t^{\bar{t}} p \tag{38}$$

$$\rho = r + \pi: \tag{39}$$

$$\left| \begin{array}{ll} -\rho_t - \alpha\rho_x - \varepsilon\rho_{xx} = 0, & (x,t) \in \Omega \times (0,\bar{t}) \\ \rho(x,\bar{t}) = r(x,\bar{t}), & x \in \Omega \\ \rho_x(0,t) = c/\varepsilon, \quad \rho(1,t) = \pi(t), & t \in (0,\bar{t}) \end{array} \right. \tag{40}$$

and it is necessary the existence of $\bar{p}$ such that $\rho(0,t) = \bar{\pi}(t) = -\int_t^{\bar{t}} \bar{p}$. Here $r(x,\bar{t})$ is as in (33). Now, deriving term by term in respect to

$t$ (this is always possible, by the regularity of the solution), (40) turns into

$$\begin{cases} - w_t - \alpha w_x - \varepsilon w_{xx} = 0 \ , & (x,t) \in \Omega \times (0,\bar{t}) \\ w(x,\bar{t}) = - \alpha r_x(x,\bar{t}) - \varepsilon r_{xx}(x,\bar{t}) - 1 \ , & x \in \Omega \\ w_x(0,\bar{t}) = c/\varepsilon, \quad w(1,t) = p(t) \ , & t \in (0,\bar{t}), \end{cases} \tag{41}$$

which can be decomposed in (30) to (33). Hence the problem is to show the existence of $\bar{p} \in H^{\frac{3}{4}}(0,\bar{t})$ $-1 \leq \bar{p} \leq 1$ such that

$$w(0,t\ ;\bar{p}) = h(t) + L\bar{p}(t) \tag{42}$$

is equal to $\bar{p}$. With this we guarantee that $\bar{q} = q(\bar{p}) = c$, for $t \in (0,\bar{t})$. But this depends on $F = h + Lp$ to have a fixed point $\bar{p}$ in $H^{3/4}(0,\bar{t})$, and, obviously, $\bar{p}$ to verify $-1 \leq \bar{p} \leq 1$.

We will prove that $F$ has a fixed point $\bar{p}$, by the usual contraction property. Before showing the contractibility of $L$ in $H^{3/4}(0,\bar{t})$, some a priori estimations are necessary. After all, as a consequence of the fixed point result, it will be demonstrated that the constraints $-1 \leq \bar{p} \leq 1$ are verified.

*Lemma 4:* Let $p \in L^\infty(0,\bar{t})$. Note $Q_{\bar{t}} = \Omega \times (0,\bar{t})$. Then, in (30) one has:

$$|y|_{L^\infty(Q_{\bar{t}})} \leq |p|_{L^\infty(0,\bar{t})} \tag{43}$$

$$\int_t^{\bar{t}} \int_0^1 \psi_1 (y_x)^2 \leq C(\alpha,\varepsilon,\psi_1)(\bar{t}-t)\ |p|^2_{L^\infty(0,\bar{t})} \ , \quad t \in (0,\bar{t}) \tag{44}$$

$$\int_0^1 \psi_2 (y_x)^2 \leq C(\alpha,\varepsilon,\psi_1,\psi_2)(\bar{t}-t)\,|p|^2_{L^\infty(0,\bar{t})} \ , \quad t \in (0,\bar{t}) \tag{45}$$

$$\int_t^{\bar{t}} \int_0^1 \psi_2 (y_t)^2 \leq D(\alpha,\varepsilon,\psi_1,\psi_2)(\bar{t}-t)\,|P|^2_{L^\infty(0,\bar{t})} \ , \quad t \in (0,\bar{t}) \tag{46}$$

$$\int_t^{\bar{t}} \int_0^1 \psi_3 (y_{xx})^2 \leq C(\alpha,\varepsilon,\psi_1,\psi_2,\psi_3)(\bar{t}-t)\,|P|^2_{L^\infty(0,\bar{t})} \ , \quad t \in (0,\bar{t}) \ , \tag{47}$$

where $\psi_i(x) = \begin{cases} 1, & x\varepsilon [0,P_i] \\ 0, & x\varepsilon [Q_i,1] \end{cases}$ $\psi_i(x) \geq 0$, $x\varepsilon[0,1]$, $\psi_i \varepsilon C^2[0,1]$, $i = 1,2,3$,

with $0<P_3<P_2<P_1<Q_3<Q_2<Q_1\leq 1$, such that $\mathrm{supp}(\psi_{i+1}) C \mathrm{supp}(\psi_i)$, $i = 1,2$, and $|\psi_2'| \leq$ constant. $(\psi_2)^{1/2}$ in $[P_2,Q_2]$ (notice that the Hermite polynomials $(x-Q_i)^2(2x+Q_i-3P_i)/(Q_i-P_i)^3$ in $(P_i,Q_i)$ verify these hypothesis).

Proof: we point out only the procedure. The details are in [7].

The estimation (43) is obtained by substituting y by $|p|_{L^\infty(0,\bar{t})}-Z$ in system (30); following this, multiply the resulting equation in Z by $Z^-$, the negative part of $Z = Z^+ + Z^-$. One gets $Z \geq 0$, therefore $y \leq |\pi|_{L^\infty(0,\bar{t})}$. Analogously, with $Z - |p|_{L^\infty(0,\bar{t})}$, we arrive to the desired result.

The other estimations are obtained by multiplying the equation in (30) by convenient factors, and using the preceding estimations.

Use the factor $\psi_1 y$ for the second, $\psi_2 y_t$ for the third and fourth, and $\psi_3 y_{xx}$, for the fifth.

*Corollary 2:* L is a contraction in $H^{3/4}(\bar{t}-t,\bar{t})$, for t sufficiently close from $\bar{t}$.

Proof: suppose $p \varepsilon H^{3/4}(0,T)$ (thus p can be considered continuous). Let M be a bound for both $H^{3/4}(0,T)$ and $L^\infty(0,T)$ norms of p. Applying the lemma 4, one gets the result.

*Lemma 5:* L is a contraction in $H^{3/4}(0,\bar{t})$.

Proof. Apply lemma 4 successively in intervals $(t_i,t_{i+1})$ that recover $(0,\bar{t})$. Itis worth-while to notice that this recovering is attainable after a finite number of iterations, because the constants involved do not depend on time t not even the solution y of each partial system.

*Lemma 6:* F has a fixed point $\bar{p}$ such that $-1 \leq \bar{p} \leq 1$.

proof: the procedure is to show, initially, that $L$ is a positive operator; and then to consider Picard iteration for $P = g + LP$, for some $g \geq 0$. In this case, it is easily seen that the fixed point $\bar{P}$ is non negative. Applying this result, successively, for $g = 1 - h - L1$, $P = 1 - \bar{p}$, and $g = 1 + h - L1$, $P = 1 + \bar{p}$, it is sufficient to show that $g \geq 0$. This can be done by the application of maximum-minimum principles.

*Theorem:* Let be $f$ as in proposition 3. Then

$$\bar{u} = f \chi_{(0,\bar{t})} \tag{48}$$

is a solution of (P), for some $\bar{t} \in (0,T)$, given as is Proposition 5. If such a $\bar{t}$ doesn't exist,

$$\bar{u} = 0 \tag{49}$$

is a solution of (P)


1. Bensousam, A.; Nissen, G.; Tapiero, C.S.: Optimum inventory and product quality control with deterministic and stochastic deterioration. An application of distributed parameter control systems. IEEE Transactions of Automatic Control, Juin (1975).

2. Lions, J.L.; Magenes, E.: Problèmes aux limites non-homogènes et applications. Vol. 2. Paris: Dunod (1968).

3. Rockafellar, R.T.: Conjugate duality and optimization. Regional Conference Series in Applied Mathematics, 16, Philadelphia: SIAM. (1974).

4. Ekeland, I.; Teman, R.: Analyse Convexe et Problèmes Variationnels. Paris: Dunod (1973).

5. Rockafellar, R.T.: Integrals which are convex functionals. Pacific Journal Math. 24, Nº 3 (1968).

6. Rockafellar, R.T.: Integrals which are convex functionals II. Pacific Journal Math. 39, Nº 2 (1971).

7. Oliveira, P.R.: Contrôle de Processus de Vieillissement. (Thèse Dr. Ing.). Mathématiques de la Décision Faculté des Sciences de Paris IX (1977).

# A PERSPECTIVE ON CONTROL SYSTEM DESIGN BY MEANS OF SEMI-INFINITE OPTIMIZATION ALGORITHMS.

E. Polak
Department of Electrical Engineering and Computer Sciences
and the Electronics Research Laboratory
University of California, Berkeley, California 94720

**ABSTRACT**

We present an overview of the use of semi-infinite optimization algorithms in linear, multivariable control system design. We deal with problem formulation, basics of algorithms, numerical aspects and software requirements.

## 1. INTRODUCTION

The first step in designing a linear control system is the selection of a structure for the compensator blocks. The second step is the assignment of values to the compensator block parameters in such a way as to satisfy design specifications. The second step is usually an iterative process which can be simplified considerably by making use of parametric optimization. Until a decade ago, the designer had to make do with differentiable optimization techniques which required that the cost function and constraints be expressed as finite sets of inequalities involving differentiable functions. This fact limited considerably the extent to which the designer could ensure that specifications were satisfied.

For example, suppose that $x$ is the vector of designable compensator parameters and that the designer is required to ensure that the magnitude $|H_{yd}(x,j\omega)|$ of the disturbance to output transfer function of a SISO system is suppressed over a given frequency range, $[\omega',\omega'']$. The natural expression of this requirement has the form

$$|H_{yd}(x,j\omega)| \leq b(j\omega), \ \forall\ \omega \in [\omega', \omega'']. \tag{1.1}$$

Clearly, (1.1) is an infinite system of nondifferentiable inequalities. Hence, to convert them to an acceptable finite set of differentiable inequalities, the designer would have to discretize the frequency interval into N points $\omega_k = \omega' + k(\omega''-\omega')/N$, with $k = 0, 1, ,..., N$, and square terms in (1.1) to make them differentiable. This would result in the set of $N+1$ differentiable inequalities:

$$|H_{yd}(x,j\omega_k)|^2 - b(j\omega_k)^2 \leq 0,\ k = 0,1,,2,...,N\,. \tag{1.2}$$

The major problem with this transcription is that when $N$ is large (say, larger than 100), a large number of gradients have to be computed at each iteration of a typical constrained optimization algorithm, producing a heavy computational burden.

Now consider the requirement of disturbance suppression in the MIMO case. Since the induced euclidean norm of a matrix is its maximum singular value, which we shall denote by $\bar{\sigma}[\cdot]$, the MIMO equivalent of (1.1) is

$$\bar{\sigma}[H_{yd}(x,j\omega)] \leq b(j\omega),\ \forall\ \omega \in [\omega',\omega'']. \tag{1.3}$$

Let us assume that $H_{yd}(x,j\omega)$ is an $m \times m$ matrix. Since for any $m \times m$ complex valued matrix $H$, $\bar{\sigma}[H] = \max\{\langle v, H^T H v\rangle \mid v \in \mathbb{C}^m, \|v\| = 1\}$, to discretize (1.3), we must discretize not only the interval $[\omega',\omega'']$, but also the unit sphere in $\mathbb{C}^m$. While 100 points may be adequate to discretize reliably the interval $[\omega',\omega'']$, the reliable discretization of the unit sphere in $\mathbb{C}^m$, when, say $m > 10$, may require considerably more than $10^{10}$ points. Clearly, such a discretization results in an unacceptable computational burden. Consequently, for a long time, the only approach to multivariable control system design was the one provided by linear quadratic gaussian regulator theory (see, e.g., [Ath.1, Kwa.1, Saf.2, Ste.1]), which is a kind of penalty function technique. While it enabled designers to produce stable, reasonably well behaved designs, it was not able to ensure the satisfaction of constraints such as (1.3).

Fortunately, over the last eight years a much more powerful approach has become possible with the development of a theory of nonsmooth analysis and optimization [Cla.1] and of semi-infinite optimization algorithms for engineering design (see e.g., [Gon.1, Pol.1, Pol.2, Pol.3]). Because, on the conceptual level, semi-infinite optimization algorithms can deal with constraints, such as (1.3) without discretization, and because they are amenable to very efficient implementation, they open up totally new possibilities in control system design.

In this paper we shall revue the formulation of a control system design in the form of a semi-infinite optimization problem (see also [Bec.1]), we shall introduce the reader to the basics of semi-infinite optimization algorithm theory, and we shall discuss briefly both the numerical and software aspects of DELIGHT.MIMO, an interactive, optimization-based linear, multivariable, control system design environment, currently being completed in Berkeley.†

## 2. FORMULATION CONTROL SYSTEM DESIGN AS AN OPTIMIZATION PROBLEM

Control system design specifications are usually expressed in terms of constraints on time and frequency responses. The mathematical form of these constraints is

†The first version of DELIGHT.MIMO was described in [Pol.4]. The latest version is being completed by T-L. Wuu, a doctoral student of the author's.

$$\varphi^j(x,\alpha_k) \le 0, \forall\ \alpha_k \in A_k, \tag{2.1a}$$

where $x \in \mathbb{R}^n$ is the design vector (designable compensator parameters); $\varphi^k: \mathbb{R}^n \times \mathbb{R} \to \mathbb{R}$ are continuous functions such that $\nabla_x \varphi(\cdot,\cdot)$ exist and are continuous, and the sets $A_k \subset \mathbb{R} \subset \mathbb{R}^{n_k}$ are compact. Mostly, the sets $A_k$ are intervals of time or frequency. However, in the cases of constraints on singular values (since $\bar{\sigma}[H] = \max\{\langle z, H^*Hz\rangle \mid \|z\| = 1)$ and of parametric dynamic model uncertainty (see [Pol.6]), the sets $A_k$ will be of of higher dimension than one. Consequently, taking into account the possibility that there may be a cost to be optimized as well as simple constraints on the design parameters, we find that control system design problems transcribe into semi-infinite optimization problems of the form

$$\min\{ f(x) \mid g^j(x) \le 0,\ j = 1,2,\dots,m;\ \max_{\alpha_k \in A_k} \varphi^k(x,\alpha_k) \le 0,\ k = 1,2,\dots,l \}, \tag{2.1b}$$

where the cost function $f: \mathbb{R}^n \to \mathbb{R}$ and the simple constraint functions $g^j: \mathbb{R}^n \to \mathbb{R}$ are continuously differentiable.

We now give a simple example to illustrate the transcription of a control system design problem into an optimization problem of the form (2.1b). Consider the control system in Fig. 1, for which it is necessary to design a compensator $C$. The dynamics of the plant are given by

$$\left.\begin{aligned} \dot{z}_P &= A_P z_P + B_P u_P \\ y_P &= C_P z_P + D_P u_P \end{aligned}\right\}, \tag{2.2a}$$

the dynamics of the compensator are given by

$$\left.\begin{aligned} \dot{z}_C &= A_C z_C + B_C u_C \\ y_C &= C_C z_C + D_C u_C \end{aligned}\right\}, \tag{2.2b}$$

while the interconnection equations are given by

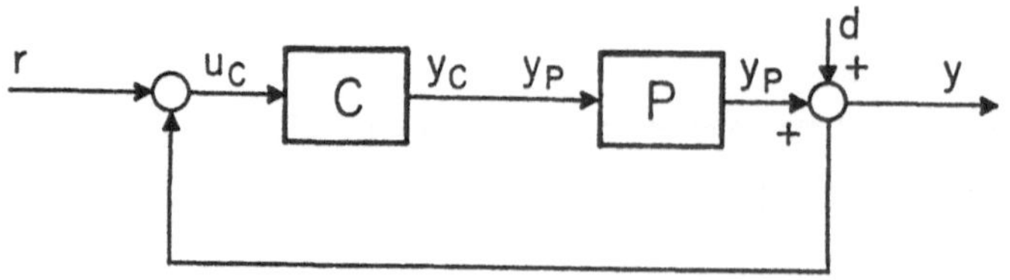


Fig. 1

$$\left.\begin{array}{l} u_P = y_C \\ u_C = -y_P - d \ + r \end{array}\right\} . \tag{2.2c}$$

The designer chooses the dimension of the compensator state vector $z_C$ and has to compute the compensator matrices $A_C$, $B_C$, $C_C$, $D_C$, whose elements eventually form the components of the design vector $x$. In order to reduce the dimension of the design vector $x$, the designer may specify the system matrix $A_C$ in block diagonal form $A_C = \mathrm{diag}(A_{1C}, A_{2C}, \ldots, A_{kC}, \lambda_{2k+1}, \ldots, \lambda_N)$, where the $\lambda_j$ are real (some may be frozen at zero for integral action), while the $2\times2$ submatrices $A_{jC}$ are of the form

$$A_{jC} = \begin{bmatrix} 0 & 1 \\ a_{1j} & a_{0j} \end{bmatrix} . \tag{2.3}$$

In addition, some structural simplification of the $B$ matrix is also possible.

Now consider typical design specifications in time and frequency domains.

(i) **Time Domain Constraints:** Given vector step inputs $r_i(t)$ such that $r_i(t) = 0$ for all $t < 0$ and $r_i(t) = (0,0,\ldots,1,0\ldots0)$ for all $t \geq 0$ (with the 1 in the i$^{th}$ place), we require that the corresponding step responses $y^i(t,x,r_j)$ remain within the envelopes shown in Fig. 2. This leads to the two semi-infinite constraints

$$y^i(x,t,r_j) - b_u^{ij}(t) \leq 0, \quad \forall\, t \in [0,T], \tag{2.4a}$$

$$-y^i(x,t,r_j) + b_l^{ij}(t) \leq 0, \quad \forall\, t \in [0,T]. \tag{2.4b}$$

Constraints on interactions are expressed by the inequalities in which $i \neq j$.

(ii) **S-plane Constraints:** When the interconnection equations (2.2c) are eliminated, we get the state equations of the closed loop system, in the form

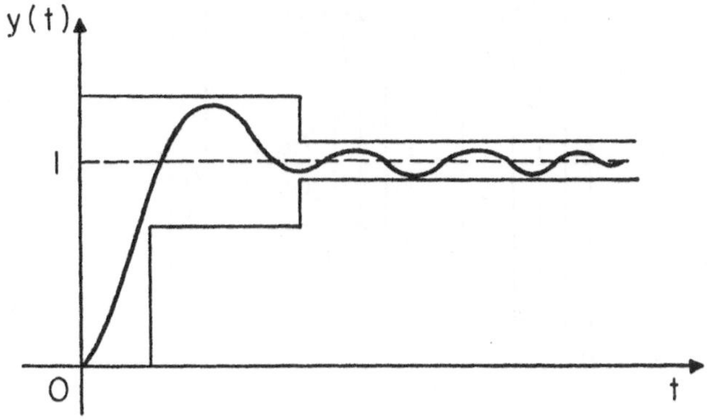


Fig. 2

$$\left.\begin{aligned} \dot{z} &= A(x)z + B(x)r \\ y &= C(x)z + D(x)r \end{aligned}\right\}, \tag{2.5a}$$

where all the matrices may depend on the design vector. The system matrix $A(x)$ of the closed loop system has the form

$$A(x) = \begin{bmatrix} A_P - B_p(I + D_C D_P)^{-1} D_C C_P & -B_P(I + D_C D_P)^{-1} C_C \\ B_C(I + D_P D_C)^{-1} C_P & A_C - B_C(I + D_P D_C)^{-1} D_P C_C \end{bmatrix}. \tag{2.5b}$$

It is common to require that all the eigenvalues of $A(x)$ lie in a sector in $\overset{o}{\mathbb{C}}_-$ (see Fig. 3). Denoting the eigenvalues of $A(x)$ by $\lambda^j[A(x)]$, the cone requirement is expressed as a system of inequalities

$$\mathrm{Im}[\lambda^j[A(x)]] + \xi[\mathrm{Re}\lambda^j[A(x)]] + \zeta \leq 0 \quad \text{for } j = 1,2,\ldots,N, \tag{2.6}$$

where $\xi$, $\zeta > 0$. Note that in (2.6) we are exploiting the fact that complex eigenvalues must occur in complex conjugate pairs. Since eigenvalues are not locally Lipschitz continuous at points of multiplicity, these inequalities must be replaced by a semi-infinite inequality expressing the modified Nyquist criterion described in [Pol.5].

(iii) **Frequency Domain Constraints:** These constraints arise from the need to suppress disturbances, saturation effects and unwanted interactions, as well as from the need to satisfy bandwidth requirements and to ensure robustness, i.e., stability in the face of plant uncertainty (see e.g., [Che.1, Des.1, Doy.1, Saf.1]). For example, when there is some unstructured uncertainty in the plant model, (2.2a) represents only the structured part $P_0$ of the plant

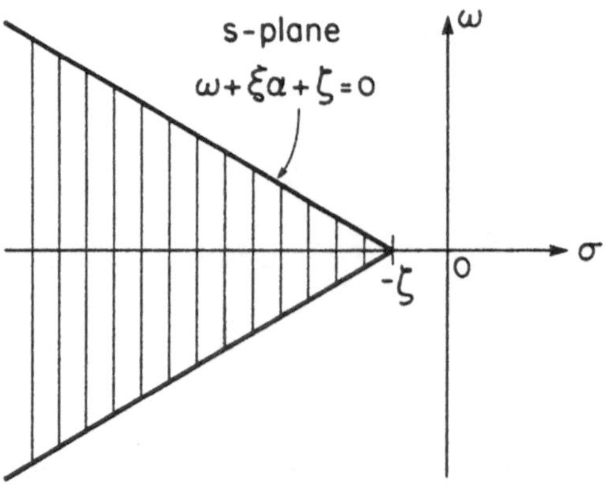


Fig. 3

model, while the actual plant has a transfer function matrix of the form $P(s) = P_0(s)(I + L(s))$, with $P_0(s)$ the transfer function matrix of (2.2a) and $L(s)$ a perturbation known only to the extent that

$$\bar{\sigma}_M[L(j\omega)] \leq b(\omega) \ \forall\ \omega \geq 0 \tag{2.7}$$

where $\bar{\sigma}[\cdot]$ denotes the largest singular value of the matrix in brackets. In that case (see [Che. 1]), for the closed loop system to be stable, we require, in addition to (2.6), that

$$\bar{\sigma}[H^0_{yr}(x,j\omega)] - \frac{1}{b(\omega)} \leq 0, \quad \forall\ \omega \geq 0, \tag{2.8}$$

where $H^0_{yr} = (I + P_0C)P_0C$. Clearly, (2.8) is a semi-infinite inequality.

(iv) **Cost Functions:** As a cost function we can take any design requirement which is not prespecified. Thus, for example we may choose to minimize the effects of a disturbance over a critical frequency range. This leads to the cost function (see [Doy. 1])

$$f(x) = \max_{\omega \in [\omega', \omega'']} \bar{\sigma}[H^0_{yd(x,j\omega)}], \tag{2.9}$$

where $H^0_{yd} = (I + P_0C)^{-1}$.

We see that when the above described constraints are supplemented with simple inequalities bounding the design variables, one obtains a transcription of the simple control system design problem into a form of problem (2.1b). Note that with the exception of the modified Nyquist inequality described in [Pol.5], the functions in the frequency domain constraints are locally Lipschitz continuous only in the subset of the design parameter space where the closed loop system is stable. Because of this, the algorithms discussed in the next section have to be modified into phase 0 - phase I - phase II versions, with phase 0 dedicated to the computation of a stabilizing controller, while in phase I and phase II, stability must be preserved throughout the entire computation.

## 3. BASICS OF SEMI-INFINITE OPTIMIZATION ALGORITHMS

The simplest problem of the form (2.1b), which captures its essence and which can be used to explain the nature of semi-infinite optimization algorithms for the solution of (2.1b) is the following one:

$$\min\{f(x) \mid \varphi(x,\alpha) \leq 0 \ \forall\ \alpha \in A\}, \tag{3.1}$$

where $f:\mathbb{R}^n \to \mathbb{R}$ is continuously differentiable, $\varphi:\mathbb{R}^n \times \mathbb{R} \to \mathbb{R}$ is locally Lipschitz continuous and $A \subset \mathbb{R}$ is a compact interval.

The semi-infinite combined phase I-phase II optimization algorithms for solving (3.1) [Gon.1, Pol.1, Pol.2, Pol.3] are obtained by extension of simple phase I-phase II methods of feasible directions (see [Pol.7, Pol.8]). Phase I - phase II methods of feasible directions compute a feasible point very rapidly. After that, they reduce the cost without violation of constraints. Since to a large measure, design consists of simply satisfying specifications, the advantage of phase I-phase II methods is clear.

The phase I-phase II algorithms described in [Pol.8] solve problems of the form

$$\min\{f(x) \mid g^j(x) \le 0,\ j \in \mathbf{m}\}, \tag{3.2}$$

where $f:\mathbb{R}^n \to \mathbb{R}$, $g^j:\mathbb{R}^n \to \mathbb{R}$, $j \in \mathbf{m}$, are continuously differentiable and $\mathbf{m} \triangleq \{1,2,\dots,m\}$. For any $x \in \mathbb{R}^n$ let

$$\psi(x) \triangleq \max_{j \in \mathbf{m}} g^j(x). \tag{3.3}$$

Then, assuming that $\min_{x \in \mathbb{R}^n} \psi(x) \le 0$, to find a point $x^*$ satisfying the constraints $g^j(x) \le 0$ one can try to solve the unconstrained (phase I) problem

$$\min_{x \in \mathbb{R}^n} \psi(x). \tag{3.4}$$

Suppose that $x^*$ is a solution of (3.4). Then (see [Cla.1]) it must satisfy the following, Kuhn-Tucker like condition of optimality:

$$\sum_1^m \mu^j \nabla g^j(x^*) = 0, \tag{3.5a}$$

$$\sum_1^m \mu^j [\psi(x^*) - g^j(x^*)] = 0, \tag{3.5b}$$

$$\sum_1^m \mu^j = 1,\ \mu^j \ge 0\ \forall\ j \in \mathbf{m}. \tag{3.5c}$$

Note that (3.5b) is a kind of complementary slackness condition. Let $\bar{G}\psi:\mathbb{R}^n \to 2^{\mathbb{R}^{n+1}}$ be defined by

$$\bar{G}\psi(x) \triangleq \mathrm{co}\left\{\begin{pmatrix} \psi(x) - g^j(x) \\ \nabla g^j(x) \end{pmatrix}\right\}_{j \in \mathbf{m}}, \tag{3.6}$$

then (3.5a-3.5b) can be expressed in the compact form $0 \in \bar{G}\psi(x^*)$. Next let

$$\bar{h}(x) = (h^0(x), h(x)) \triangleq -\mathrm{Nr}[\bar{G}\psi(x)],$$
$$= \mathrm{argmin}\{\|\bar{h}\| \mid \bar{h} \in \bar{G}\psi(x)\}. \quad (3.7)$$

Referring to (3.6), we find that $\bar{G}\psi(\cdot)$ is continuous and hence, from (3.7), that $\bar{h}(\cdot)$ is continuous. Furthermore, the directional derivative of $\psi$ at $x$, in the direction $h(x)$, satisfies

$$d\psi(x; h(x)) \leq -\|\bar{h}(x)\|^2, \quad (3.8)$$

Combining these observations, we conclude that $h(x)$ is a continuous descent direction for the phase I problem (3.4). Consequently, one can solve (3.4) by making use of the following simple generalization of the Armijo method [Arm.1] (see [Pol.3]):

**Algorithm 3.1.**

Parameters:
$\alpha, \beta, \in (0,1)$.

Data: $x_0 \in \mathbb{R}^n$.

Step 0: Set i = 0.

Step 1: Compute the augmented search direction $\bar{h}_i = (h^{0_i}, h_i) = -\mathrm{Nr}[\bar{G}\psi(x_i)]$ and extract the actual search direction $h_i$.

Step 2: Compute the step size $\lambda_i$, according to the rule

$$\lambda_i = \underset{k \in \mathbb{N}}{\mathrm{argmax}}\{\beta^k \mid \psi(x_i + \beta^k h_i) - \psi(x_i) \leq -\beta^k \alpha \|\bar{h}_i\|^2\}.$$

Step 3: Update: set $x_{i+1} = x_i + \lambda_i h_i$, replace $i$ by $i + 1$ and go to step 1. ■

**Theorem 3.1**[Pol.3]: If $x^*$ is an accumulation point of $\{x_i\}_{i=0}^{\infty}$ constructed by Algorithm 3.1, then $0 \in \bar{G}\psi(x^*)$. ■

Hence, if $0 \notin \bar{G}\psi(x)$ for all $x \in \mathbb{R}^n$ such that $\psi(x) \geq 0$ and $\min_{x \in \mathbb{R}^n} \psi(x) < 0$, then Algorithm 3.1 will compute a point $x_k$ satisfying $\psi(x_k) \leq 0$ in a finite number of iterations.

To extend Algorithm 3.1 to the case where $\psi(x) \triangleq \max_{y \in Y} \varphi(x,y)$ we only need to redefine $\bar{G}\psi(x)$ to be (see [Pol.3])

$$\bar{G}\psi(x) \triangleq co\left\{\begin{bmatrix}\psi(x)-\varphi(x,\alpha)\\ \nabla\psi(x,\alpha)\end{bmatrix}\right\}_{\alpha \in A}. \tag{3.9}$$

Next, to extend Algorithm 3.1 to the case of Problem (3.2), we note that the F. John first order optimality conditions for (3.2) can be expressed in the form

$$\mu^0 \nabla f(\hat{x}) + \sum_{j=1}^{m} \mu^j \nabla g^j(\hat{x}) = 0, \tag{3.10a}$$

$$\sum_{j=1}^{m} \mu^j g^j(\hat{x}) = 0, \tag{3.10b}$$

$$\sum_{j=0}^{m} \mu^j = 1, \ \mu^j \geq 0 \ \forall \ j \in \underline{m+1}. \tag{3.10c}$$

This suggests the following definition of the *phase II search direction finding map* $\bar{G}f_{II}{}^{,\psi} : \mathbb{R}^n \to 2^{\mathbb{R}^{n+1}}$ (see [Pol.3])

$$\bar{G}f_{II}{}^{,\psi}(x) \triangleq co\left\{\begin{bmatrix}0\\ \nabla f(x)\end{bmatrix}, \begin{bmatrix}-g^j(x)\\ \nabla g^j(x)\end{bmatrix}\right\}_{j \in \underline{m}}. \tag{3.11}$$

Then (3.10a-3.10c) can be expressed in the compact form $0 \in \bar{G}f_{II}{}^{,\psi}(\hat{x})$. Next let

$$\bar{h}(x) = (h^0(x), h(x)) \triangleq -\mathrm{Nr}[\bar{G}f_{II}{}^{,\psi}(x)],$$

$$= \mathrm{argmin}\{\|\bar{h}\| \mid \bar{h} \in \bar{G}f_{II}{}^{,\psi}(x)\}. \tag{3.12}$$

Referring to (3.11), we find that $\bar{G}f_{II}{}^{,\psi}(\cdot)$ is continuous and hence, from (3.12), that $\bar{h}(\cdot)$ is continuous. Furthermore, the directional derivatives of $f$ and $\psi$ satisfy

$$df(x;h(x)) \leq -\|\bar{h}(x)\|^2, \tag{3.13a}$$

$$h^0(x)\psi(x) + d\psi(x;h(x)) \leq -\|\bar{h}(x)\|^2, \tag{3.13b}$$

Combining these observations, we conclude that $h(x)$ is a continuous feasible descent direction for the phase II part of problem (3.2) (i.e., for solving (3.2) given an $x_0$ such that $\psi(x_0) \leq 0$. Consequently, one can solve (3.2) by making use of the following method see [Pol.3]):

**Algorithm 3.2**[Pol.3]:

Parameters:

$\alpha, \beta, \gamma \in (0,1), \gamma > 0.$

Data: $x_0 \in \mathbb{R}^n$ such that $\psi(x_0) \leq 0$.

Step 0: Set i = 0.

Step 1: Compute the augmented search direction $\overline{h}_i = (h_i^0, h_i) = -\mathrm{Nr}[\overline{G}f_I{}^{,\psi}(x_i)]$ and extract the actual descent direction $h_i$.

Step 2: Compute the step size $\lambda_i$, according to the rule

$$\lambda_i = \arg\max_{k \in \mathbb{N}} \{\beta^k \mid f(x_i + \beta^k h_i) - f(x_i) \leq -\beta^k \alpha \|\overline{h}_i\|^2);\ \psi(x_i + \beta^k h_i) \leq 0\} \tag{3.14}$$

Step 3: Update: set $x_{i+1} = x_i + \lambda_i h_i$, replace $i$ by $i + 1$ and go to step 1. ▪

**Theorem 3.2:** If $\hat{x}$ is an accumulation point of a sequence $\{x_i\}_{i=0}^{\infty}$ constructed by Algorithm 3.2, then $\psi(\hat{x}) \leq 0$ and $0 \in \overline{G}f_I{}^{,\psi}(\hat{x})$, i.e., $\hat{x}$ satisfies the generalized F. John condition of optimality (see [Cla.1]). ▪

To extend Algorithm 3.2 to the phase II part of problem (3.1), we define $\psi(x) \triangleq \max_{\alpha \in A} \varphi(x, alpha)$, we only need to redefine $\overline{G}f_I{}^{,\psi}(x)$ to be (see [Pol.3])

$$\overline{G}f_I{}^{,\psi}(x) \triangleq co\left\{\begin{bmatrix} 0 \\ \nabla f(x) \end{bmatrix}, \begin{bmatrix} -\varphi(x,\alpha) \\ \nabla\psi(x,\alpha) \end{bmatrix}\right\}_{\alpha \in A}. \tag{3.15}$$

It is desirable to combine the phase I and phase II computations into a single procedure with the property that when $\psi(x)>0$ is large, it behaves essentially as the phase I method, while when $\psi(x) \leq 0$, it reduces to the phase II method. In the transition phase, the desire to reduce the cost should be allowed to become progressively more dominant as feasibility is approached. This can be achieved by introducing a phase I - phase II search direction finding map [Pol.3], as follows:

$$\overline{G}^{f,\psi}(x) \triangleq co\left\{\begin{bmatrix} \gamma\psi(x)_+ \\ \nabla f(x) \end{bmatrix}, \begin{bmatrix} \psi(x)_+ - \varphi(x,\alpha) \\ \nabla\psi(x,\alpha) \end{bmatrix}\right\}_{\alpha \in A}. \tag{3.16}$$

where $\psi(x)_+ \triangleq \max\{\psi(x), 0\}$ and $\gamma > 0$ is arbitrary. This latest search direction finding map leads to the following phase I -phase II method:

**Algorithm 3.3:**

Parameters:

$\alpha, \beta \in (0,1)$, $\gamma > 0$.

Data: $x_0 \in \mathbb{R}^n$.

Step 0: Set i = 0.

Step 1: Compute the augmented search direction $\bar{h}_i = (h_i^0, h_i) = -\mathrm{Nr}[\bar{G}^{f,\psi}(x_i)]$ and extract the actual descent direction $h_i$.

Step 2: Compute the step size $\lambda_i$, according to the rule

$$\lambda_i = \arg\max_{k \in \mathbb{N}} \{\beta^k \mid \psi(x_i + \beta^k h_i) - \psi(x_i) \le -\beta^k \alpha \|\bar{h}_i\|^2)\} \tag{3.17}$$

if $\psi(x_i) > 0$,

$$\lambda_i = \arg\max_{k \in \mathbb{N}} \{\beta^k \mid f(x_i + \beta^k h_i) - f(x_i) \le -\beta^k \alpha \|\bar{h}_i\|^2); \; \psi(x_i + \beta^k h_i) \le 0\}$$

if $\psi(x_i) \le 0$.

Step 3: Update: set $x_{i+1} = x_i + \lambda_i h_i$, replace $i$ by $i + 1$ and go to step 1. ∎

**Theorem 3.3**[Pol.3]: Suppose that $0 \notin \bar{G}\psi(x)$ (defined in (3.9) for all $x \in \mathbb{R}^n$ such that $\psi(x) \ge 0$. If $\hat{x}$ is an accumulation point of any sequence $\{x_i\}_{i=0}^{\infty}$ constructed by Algorithm 3.3, then $\psi(\hat{x}) \le 0$ and $0 \in \bar{G}f_{ff}^{,\psi}(\hat{x})$ (i.e., $\hat{x}$ satisfies the generalized F. John condition of optimality (see [Cla.1])).

This completes our elementary exposition of the ideas governing the construction of semi-infinite algorithms. For an in depth treatment of this topic, we refer the reader to [Pol.3]. In particular, the reader will find in [Pol.3] algorithms in which the search direction is computed using only $\varepsilon$-active constraints. Such search directions are not continuous and hence algorithms using them are difficult to analyze. However, $\varepsilon$-active search directions are much cheaper to compute than the ones discussed above. In addition, the reader will find in [Pol.3, Pol.6] details of applications to control system design problems with singular value constraints or optimal control requirements.

## 4. INTERACTIVE SOFTWARE FOR OPTIMIZATION-BASED CONTROL SYSTEM DESIGN

Optimization-based control system design requires highly specialized software. To our knowledge, there is currently only one interactive, optimization-based, control system design software system being developed. This system is called DELIGHT.MIMO (see [Pol.4]) and it is being developed jointly by research teams at the University of California, Berkeley, and Imperial College, London.

DELIGHT.MIMO is a member of a family of optimization-based CAD packages implemented in the DELIGHT system [Nye.1, Nye.2]. Hence a description of DELIGHT.MIMO must begin with a brief description of DELIGHT which can be thought of as a highly portable operating system for a FORTRAN or C machine. As can be expected from an operating system, it provides a certain number of commonly found features such as a text editor, a read and write files command, an ability to install and execute FORTRAN and C programs, a help command, a history command, a repeat command, hard interrupts, etc.

In addition, DELIGHT provides a number of rather special features. The most important of these are the following.

(i) **Color graphics commands** for interaction with data and programs. The graphics commands allow the user to define viewports and windows and two produce displays by making use of elementary operations such as vector, move, cursor and text. These are used when display improvisation is necessary. In addition there are also a number of high level commands of the form plot data according to the options specified, which can be used to produce various orthodox type plots.

(ii) **RATTLE, a high level language**, for the programming of optimization algorithms as well as information display options. RATTLE requires about 1/10 of the number of program lines compared to FORTRAN, but it executes considerably slower that FORTRAN. Because of this, in design packages a mixture of RATTLE, C and FORTRAN code is always used.

RATTLE compiles incrementally, it has binary matrix operation capability, it uses defines for command simplification and macros for producing simple RATTLE calls to complex FORTRAN programs (e.g. linprog x = argmin {<c,z>| Az=0, Bz>= 0, z>=0). It is easy to use RATTLE to construct code for conversational data entry.

(iii) **Interactive debugging and executing program modification tools.** These consist of soft interrupts, an enter command and a reset command. Unlike hard interrupts which suspend a program the instant the break key is depressed, soft interrupts suspend a program only at designated break points in the program. When either a hard or a soft interrupt is executed, it is possible to enter suspended subprocedures and display and modify both local and global variables. It is also possible to modify the code defining the problem being solved as well as the code defining the algorithm. After an interrupt the user may start up a totally unrelated computation or resume execution of the suspended program. To return to a suspended program, the user executes the the reset command.

(iv) **A structure for a modular, RATTLE code, optimization algorithm library.** This library enables the user to assemble an algorithm from optional blocks, such as step size and direction finding procedures, via a menu. The problem to be solved must be described by means of several files containing either dimensional information or RATTLE code for: the cost function, ordinary inequality constraints, functional inequality constraints, and gradients of the appropriate functions. The optimization problem and algorithm are compiled and linked by means of the solve command, e.g., ***solve pid using Polak-Wardi***, when neither the problem pid nor algorithm Polak-Wardi has been compiled, or ***solve pid*** (or ***solve using Polak-Wardi)*** when the algorithm (problem) have been compiled earlier. Algorithms can be executed a desired number of iterations by means of the ***run k*** command, or they can be executed one step at a time, by means of the ***step k*** command. When execution of an optimization program is interrupted by means of a soft or hard interrupt, the user may adjust algorithm parameters, completely replace the algorithm, modify the problem description files, display variable values or plot response graphs.

DELIGHT.MIMO adds to the basic DELIGHT system a sophisticated database for control system, bottom up and top down interconnection description for subsystems in state form, programs for control system time and frequency response simulation, a symbolic differentiator for obtaining derivatives of parametrized system matrices parameters, interactive programs for initial design generation, an interactive program which assists the user in forming the problem description files from design specifications, as required by the optimization algorithm library format, and both alpha numeric and graphical means for entering the control system configuration. The optimization algorithm currently used for control system design is the Polak-Wardi method described in [Pol.2]. DELIGHT.MIMO has its own help facility which is treelike in structure.

An important aspect of the MIMO part of DELIGHT.MIMO is the graphical interaction facility. It is mostly commanded by means of a mouse, though some data, such as component names must be entered via the keyboard. It includes a block diagram editor, windowing capability, main and pop menus, various graphical aids for entering design specifications and cost function, sophisticated display capability for system responses, and a graphical display of system response sensitivities. This graphical interface makes the whole system a very accessible and flexible design tool.

## 5. COMPUTATION OF SYSTEM RESPONSES AND THEIR DERIVATIVES

A control system is invariably an interconnection of subsystems. When these subsystems are represented in state space form, as

$$S_i: \quad \left.\begin{aligned} \dot{x}_i &= A_i x_i + B_i u_i \\ y_i &= C_i x_i + D_i u_i \end{aligned}\right\}, \tag{5.1}$$

the equations for the overall system $S$ can be written in the form

$$S: \quad \left. \begin{aligned} \dot{z} &= Az + Bu \\ y &= Cz + Du \end{aligned} \right\}, \tag{5.2}$$

$$u = Ey + Jr , \tag{5.3}$$

where, in (5.2), assuming that there are $N$ subsystems, $A = diag(A_1, A_2, \dots, A_N)$, $B = diag(B_1, B_2, \dots, B_N)$, $C = diag(C_1, C_2, \dots, C_N)$, $D = diag(D_1, D_2, \dots, D_N)$ and (5.3) expresses the interconnections between the subsystems (algebraically). In (5.3), r is a vector of external inputs and E and J are matrices whose elements are zeros and ones.

The matrices $A_i$, $B_i$, $C_i$, $D_i$, specifying the subsystems may be given either in numerical form or in parametric form. In the DELIGHT.MIMO system, because of a constraint imposed by the symbolic differentiator, when given in parametric form, the elements of these matrices must be multinomials in the elements of the design parameter vector x.

When the interconnection equation (5.3) is eliminated, we obtain a reduced description of the form

$$S_c \quad \left. \begin{aligned} \dot{z} &= A_c z + B_c r \\ y &= C_c z + D_c r \end{aligned} \right\}, \tag{5.4}$$

where, in terms of the matrices in (5.2)and (5.3), $A_c = A + B[I-ED]^{-1}EC$, $B_c = B[I-ED]^{-1}J$, $C_c = C + D[I-ED]^{-1}EC$, and $D = D[I-ED]^{-1}J$.

Since the closed loop system (5.4) always has distinct eigenvalues (at least with probability 1), the computation of responses and their sensitivities to design parameter variations can be considerably simplified by diagonalization. Thus, rewriting (5.2) with the parameters made explicit, we get

$$S: \quad \left. \begin{aligned} \dot{z}(t,x) &= A_c(x)z(t,x) + B_c(x)r(t) \\ y(t,x) &= C_c(x)z(t,x) + D_c(x)r(t) \end{aligned} \right\}. \tag{5.5}$$

Assuming that $A_c(x)$ is an $m \times m$ matrix, let $\Lambda(x) = diag(\lambda_1(x), \dots, \lambda_m(x))$ be a diagonal matrix of eigenvalues of $A_c(x)$ and let $W(x)$ be a corresponding matrix of eigenvectors of $A_c(x)$. Then we have the explicit formula

$$z(t,x) = W(x)e^{\Lambda(x)t}W(x)^{-1}z(0) + \int_0^t W(x)e^{\Lambda(x)(t-s)}W(x)r(s)ds . \tag{5.6}$$

*Assuming that the input $r(t)$ is a polynomial function of $t$*, the integral in (5.6) can easily be computed by formula. The output $y(t,x)$ can then computed according to (5.5b).

Next, the derivatives of $A_c(x)$, $B_c(x)$, $C_c(x)$, $D_c(x)$, with respect to a design parameter $x^i$ are easily computed, making use of the observation that

$(\partial/\partial x^i)(I-ED(x))^{-1} = (I-ED(x))^{-1}[(\partial/\partial x^i)D(x)](I-ED(x))^{-1}$. Hence state sensitivities can be computed by solving the set of differential equations, below, in conjunction with those in (5.5a):

$$\frac{\partial}{dt}\frac{\partial z(t,x)}{\partial x^i} = A_c(x)\frac{\partial z(t,x)}{\partial x^i} + \frac{\partial A_c(x)}{\partial x^i} z(t,x) + \frac{\partial B(x)}{\partial x^i} r(t), \tag{5.7a}$$

and the output sensitivities are then given by

$$\frac{\partial y(t,x)}{\partial x^i} = C_c(x)\frac{\partial z(t,x)}{\partial x^i} + \frac{\partial C_c(x)}{\partial x^i} z(t,x) + \frac{\partial D(x)}{\partial x^i} r(t). \tag{5.7b}$$

Clearly, the solution of these is also facilitated by diagonalization of the matrix $A_c(x)$.

A computationally more efficient method, based on Li brackets, was proposed in [Wuu.1]. This method is based on an efficient technique, also based on the diagonalization of the matrix $A_c(x)$, for the evaluation of state response sensitivities from the formula

$$\begin{aligned}\frac{\partial z(t,x)}{\partial x^i} = \frac{\partial e^{A_c(x)t}}{\partial x_i} z(0) + \int_0^t \{ \frac{\partial e^{A_c(x)(t-s)}}{\partial x_i} B(x) r(s) \\ + e^{A_c(x)(t-s)} \frac{\partial B(x)}{\partial x^i} r(s)\} ds\end{aligned} \tag{5.8}$$

Next we turn to the frequency response of the interconnected system. The input output transfer function of the interconnected system is given by

$$G_c(j\omega,x) = C_c(x)(j\omega - A_c(x))^{-1}B_c(x) + D_c(x). \tag{5.9}$$

Since the derivative of $G_c(j\omega,x)$, with respect to the design parameter $x$ is not a matrix, it is easiest to obtain expressions for it componentwise, as follows,

$$\begin{aligned}\frac{\partial G_c(j\omega,x)}{\partial x_i} = \frac{\partial C_c(x)}{\partial x^i}[j\omega I - A_c(x)]^{-1}B_c(x) + D_c(x) \\ + C_c(x)[j\omega I - A_c(x)]^{-1}\frac{\partial A_c(x)}{\partial x^i}[j\omega I - A_c(x)]^{-1}B_c(x) \\ + C_c(x)[j\omega I - A_c(x)]^{-1}\frac{\partial B_c(x)}{\partial x^i} + \frac{\partial D(x)}{\partial x^i}.\end{aligned} \tag{5.10}$$

Assuming that formulae for the derivatives of the matrices $A_c(x)$,, $B_c(x)$, $C_c(x)$, $D_c(x)$ are computed by a symbolic differentiator, the only major computation in the evaluation of the frequency responses and their derivatives consists of the evaluation of the matrix $[j\omega I - A_c(x)]^{-1}$ for a large number of frequencies $\omega$. Again assuming that the matrix $A_c(x)$

diagonalizable, the evaluation of $[j\omega I - A_c(x)]^{-1}$ can be considerably simplified by making use of the formula

$$[j\omega I - A_c(x)]^{-1} = W(x)[j\omega I - \Lambda(x)]^{-1} W(x)^{-1}, \tag{5.11}$$

where $\Lambda(x)$, $W(x)$ are a diagonal matrix of eigenvalues and a matrix of corresponding eigenvectors, respectively.

## 6. DESIGN INITIALIZATION TECHNIQUES

The use of optimization in control system design is not a totally automatic process. There are several stages that require judgement and the knowledge of both optimization and control theory. For example, as we have already seen, first of all, the designer is required to transcribe design specifications into well conditioned semi-infinite inequalities. This is by no means a unique process. Next, the designer is required to decide on an initial compensator configuration and to produce a set of initial values for the compensator parameters. This is a creative process that can benefit considerably from the availability of a knowledge-based system which enables the designer to make use of the more popular nonoptimal design techniques, for example, such as those described in [Des.1, Doy.1, Kwa.1, Mac.1, Moo.1, Ros.1, Saf.1, Saf.2, Ste.1]. At the present time, the DELIGHT.MIMO system does not incorporate a knowledge base and, in general, it relies on initial designs produced by other design packages. The only exception to this is the presence of software in DELIGHT.MIMO for LQG and LQR compensator design, as well as some software implementing model reduction algorithms. In the simplest case, model reduction algorithms replace the observer dynamics with its DC gain matrix.

It should be noted that in order to introduce into the design integrators for the elimination of steady state errors, a certain amount of ingenuity must be exercised in using LQG techniques. For example, consider the case in Fig.4, where we propose to use LQR techniques to design a compensator consisting of an integrator block and a state feedback matrix. The state feedback is later to be implemented by means of an observer and the whole design simplified by means of order reduction techniques. For the purpose of designing the state feedback matrix $K_P$, the external input $r$ must be neglected and the plant input is used as the feedback point.

Now, suppose that the plant has dynamics given by (2.2a) and that the integrator of the compensating block has dynamics

$$\begin{aligned} \dot{z}_C &= u_C \\ y_C &= z_C , \end{aligned} \tag{6.1}$$

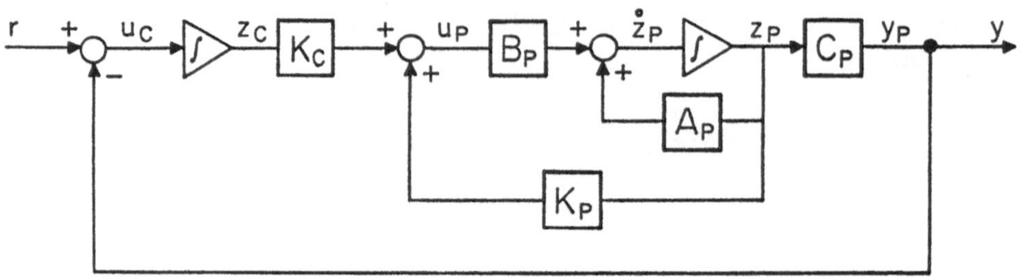


Fig. 4

with the interconnection is specified by

$$u_C = -y_P . \tag{6.2}$$

Then, for the purpose of LQR design, we must use the combined dynamics

$$\frac{d}{dt}\begin{bmatrix} z_P \\ z_C \end{bmatrix} = \begin{bmatrix} A_P & 0 \\ -C_P & 0 \end{bmatrix} + \begin{bmatrix} B_P \\ 0 \end{bmatrix} u_P . \tag{6.3}$$

LQR techniques yield a state feedback matrix $K = [K_P \mid K_C]$ which results in the feedback law

$$u_P = K_P + K_C z_C . \tag{6.4}$$

Such a design may have reasonably good transient response, but most likely, it will not satisfy most of the constraints that were discussed in Section 2. However, since it yields a stabilizing compensator, it serves as a good starting point for phase I of the algorithms that we have discussed in Section 3.

## 7. CONCLUSION

It should be clear from our presentation that the basics of the use of semi-infinite optimization algorithms in control system design are quite well understood at present. However, a great deal of work remains to be done. This includes the development of theoretical tools, such as computationally efficient parametrizations of controllers, as well as the development of software tools, such as knowledge-based systems for assisting the user in scaling problems, adjusting algorithms and performing trade offs.

**ACKNOWLEDGEMENT**

The research reported herein was sponsored in part by the National Science Foundation under Grant ECS-8121149, the Air Force Office of Scientific Research under Grant AFOSR-83-0361, the Office of Naval Research under Grant N00014-83-K-0602, the Semiconductor Research Consortium under Grant SRC-82-11-008, the State of California MICRO Program and the General Electric Co.

**8. REFERENCES**

[Arm.1] Armijo, L., "Minimization of functions having Lipschitz continuous first partial derivatives", *Pacific Journal of Mathematics* Vol 16, pp 1-3, 1966.

[Ath.1] Athans, M., "The role and use of stochastic linear-quadratic-gaussian problem in control system design", *IEEE Trans. on Automatic Control* vol. AC-16, no. 6, 1971.

[Bec.1] Becker, R. G., Heunis, A. J., and Mayne, D. Q., "Computer-Aided Design of Control Systems via Optimization", *Proc. IEE*, vol. 126, no. 6, 1979.

[Che.1] Chen, M. J. and Desoer, C. A., "Necessary and Sufficient Conditions for Robust Stability of Linear Distributed Feedback Systems", *Int. Journal on Control*, Vol. 35, No. 2, pp 255-267, 1982.

[Cla.1] Clarke, F. H., *Optimization and Nonsmooth Analysis*, Wiley-Interscience, New York, N.Y., 1983.

[Des.1] Desoer, C. A., and Gustafson C. L., "Algebraic Theory of Linear Multivariable Feedback Systems," *IEEE Trans. on Automatic Control*, VOl. AC-29, No. 10, pp. 909-917, October 1984.

[Doy.1]

Doyle, J. C., and Stein, G. "Multivariable Feedback Design: Concepts for a Classical/Modern Synthesis", *IEEE Trans. on Automatic Control*, Vol. AC-26, No. 1, pp. 4-16, 1981. Academic Press, NY, 1963.

[Gon.1] Gonzaga, C., Polak, E., and Trahan, R., "An Improved Algorithm for Optimization Problems with Functional Inequality Constraints", *IEEE Trans. on Automatic Control*, Vol. AC-25, No. 1, 1980.

[Kwa.1] Kwakernaak, H and R. Sivan, *Linear Optimal Control Systems*, Wiley-Interscience, New York, 1972.

[Mac.1] MacFarlane, A. G. J. and Postlethwaite, I., "Generalized Nyquist Stability Criterion and Multivariable Root Loci", *Int. J. Control*, Vol. 25(1) 1977.

[Moo.1] Moore, B. C., "Principal component analysis in linear systems: controllability, observability and model reduction", *IEEE Trans. on Automatic Control*, Vol. AC-26, No.1 pp. 17-32, 1981.

[Nye.1] Nye, W.T., Polak, E., Sangiovanni-Vincentelli, A., and Tits, A., "DELIGHT: an Optimization-Based Computer-Aided-Design System" *Proc. IEEE Int. Symp. on Circuits and Systems*, Chicago, Ill, April 24-27, 1981.

[Nye.2] Nye, W. T., "DELIGHT: An interactive system for optimization-based engineering design", Electronics Research Laboratory, University of California, Berkeley, Memo No. UCB/ERL M83/33, May 31, 1983.

[Pol.1] Polak, E. and Mayne, D. Q., "An Algorithm for Optimization Problems with Functional Inequality Constraints", *IEEE Trans. on Automatic Control*, Vol. AC-21, No. 2, 1976.

[Pol.2] Polak, E., and Wardi, Y. Y., "A nondifferentiable optimization algorithm for the design of control systems subject to singular value inequalities over a frequency range", *Automatica*, Vol. 18, No. 3, pp. 267-283, 1982.

[Pol.3] Polak. E., "On the mathematical foundations of nondifferentiable optimization in engineering design", University of California, Electronics Research Laboratory Memorandum No. UCB/ERL M85/17, 2/28/85. *SIAM Review*, to appear.

[Pol.4] Polak, E., Siegel P., Wuu, T., Nye, W. T., and Mayne, D. Q., "DELIGHT-MIMO an interactive, optimization based multivariable control system design package", IEEE Control Systems Magazine, Vol.2, No.4, Dec. 1982, pp 9-14.

[Pol.5] Polak, E., "A Modified Nyquist Stability Criterion for Use in Computer-Aided Design", University of California, Electronics Research Laboratory Memorandum No. UCB/ERL M83/11, *IEEE Trans. on Automatic Control*, Vol. AC-29, No. 1, pp 91-93, 1984.

[Pol.6] Polak, E., D. Q. Mayne and D. M. Stimler, "Control System Design via Semi-Infinite Optimization", *IEEE Proceedings*, pp 1777, Dec. 1984.

[Pol.7] Polak, E., *Computational Methods in optimization: A Unified Approach*, Academic Press, N.Y., 1971.

[Pol.8] Polak, E., Trahan, R., and Mayne, D. Q., "Combined phase I - phase II methods of feasible directions", *Math. Programming*, Vol. 17, No. 1, 1979, pp 32-61.

[Ros.1] Rosenbrock, H. H., *Computer-Aided Control System Design*, Academic Press, London, 1974.

[Saf.1] Safonov, M. G., Laub, A. J., and Hartman, G. L., "Feedback Properties of Multivariable Systems: The Role and Use of the Return Difference Matrix", *IEEE Trans. on Automatic Control*, Vol. AC-26, 1981.

[Saf.2] Safonov, M. G., "Choice of quadratic cost and noise matrices and the feedback properties of multiloop LQG regulators", *Proc. Asilomar Conf. on Circuits, Systems, and Computers*, Pacific Grove, California, 1979. Vol. 15, 1979.

[Ste.1] Stein, G., "Generalized quadratic weights for asymptotic regulator properties", *IEEE Trans. on Automatic Control*, Vol. AC-24, 1979.

[Wuu.1] Wuu, T. L., R. G. Becker and E. Polak, "A diagonalization technique for the computation of sensitivity functions of linear time-invariant systems," University of California, Berkeley, Electronics Research Laboratory Memo No. UCB/ERL M86/13, 14 February 1986.

# A DYNAMIC MULTIGRID ALGORITHM SUITABLE FOR PARTIAL DIFFERENTIAL EQUATIONS WITH SINGULAR SOLUTIONS.

MARIA CECILIA RIVARA
Departamento de Matemáticas y Ciencias de la Computación
Facultad de Ciencias Físicas y Matemáticas
Universidad de Chile
Casilla 170/3-Correo 3, Santiago de Chile

SUMMARY.

A dynamic multigrid algorithm able to solve systems of linear equations associated with partial differential equations having singular exact solutions is presented and discussed. This algorithm is defined in the context an adaptive finite element method over triangular meshes. The adaptive procedure provides information that assists in deciding which elements should be refined in the next grid. Furthermore, it also defines in a natural way subregions where more iterative sweeps are needed in order to preserve the optimality of the multigrid method with respect to the computational work involved. The dynamic multigrid algorithm is essentially a modified full multigrid algorithm where local or global iterative sweeps are performed between consecutive grids, depending on the number of nodes adaptively created between consecutive levels. We first summarize the construction of the non-uniform finite element hierarchy and introduce the conceps of local subproblems and submeshes. Finally we discuss the dynamic multigrid algorithm. Numerical experiments are presented and some remarks concerning research in progress are included.

## 1. INTRODUCTION.

An essential advantage of finite element discretizations is the possibility of using non-uniform meshes to manage partial differential equations with singular exact solutions. For this kind of problems an optimal discretization will be certainly irregular (more refined around the singularities). From the point of view of the implementation of a software which exploits this idea, two interesting and related techniques have been recently developed: adaptivity of the grid and the multigrid or multilevel methods for solving sparse systems of equations associated with partial differential equations. Both techniques consider that the continuous boundary value problem can be adequately approximated by a sequence of irregular discretized problems.

In [10,11], the design and implementation of a software which allows the adaptive construction of quasi-optimal sequences of finite element solutions over irregular triangulations have been extensively discussed. This software combines and adaptive strategy based on the ideas of Babuska and Rheinboldt [1-4], adequate conforming mesh refinement algorithms for triangulations [7,8] and a dynamic multigrid algorithm [10].

As stated in [9,10], adaptivity in the presence of severe singularities can produce a sequence of consecutive finite element spaces of slowly increasing dimension, and consequently, the efficiency of the classical full multigrid algorithm can be seriously affected. To deal with this problem, an adaptive multigrid algorithm, which corresponds to a modified full multigrid algorithm [5,10], was developed. The two most distinguished features of this algorithm are the following [9-11]:

(i) The finite element problem at level k+1 is solved by using a dynamic multigrid iteration only if the number of nodes adaptively created between levels k and k+1 is big enough with respect to the number of nodes of triangulation k. Otherwise, a fixed number of iterative sweeps is performed over an associated local subproblem in order to obtain an approximation of the finite element solution at level k.

ii) The dynamic multigrid iteration essentially corresponds to a V-cycle of the classical multigrid algorithm [5,10] where local or global iterative sweeps are performed between consecutive grids, depending on the number of nodes adaptively generated between both levels.

It is worth pointing out here that the development of the conforming mesh refinement algorithms has solved an essential practical question: the construction of hierarchies of non-uniform triangular finite element spaces to be used in multigrid iterations [9,10].

This paper is mainly concerned with the description of the dynamic multigrid algorithm. To this end we first assume we have at our disposal a suitable adaptive procedure and briefly discuss the procedure that allows the construction of the hierarchy of the finite element spaces. We then introduce the concepts of local subproblems and submeshes, and finally discuss the dynamic multigrid algorithm. Numerical results are presented, followed by some remarks.

## 2. CONSTRUCTION OF A HIERARCHY OF TRIANGULATIONS.

In order to use multigrid methods for partial differential equations with singular exact solutions, it is necessary to construct a hierarchy of finite element spaces $\{K_k\}$ such that $H_{k-1} \subset H_k$, and at some level k, $H_k$ has the desired non-uniform structure. The refinement process described by Rivara [7,8] allows the construction of sequences of triangulations which guarantee these properties. This is based on the bisection of triangles by the longest side and can be described as follows: given a conforming triangulation $\tau$ (where the intersection of non-disjoint triangles is either a vertex or a common side), and a refinement submesh $V \subset \tau$ (defined throughout an adaptive procedure), a new refined conforming triangulation $\tau^*$ is obtained as follows:

1. Input : $\tau_o$, V

2. $\tilde{\tau} \leftarrow \tau_o$; $j \leftarrow 0$ ;

3. For all $t \in V$ do
   Bisect t by the longest side generating $t_1, t_2$
   $\tilde{\tau} := \tilde{\tau} \setminus \{t\} \cup \{t_1, t_2\}$

4. Let S be the set of non-conforming triangles of $\tilde{\tau}$ ($t \in \tau$ is non-conforming if there exists a midpoint P of one of the sides of t which is a vertex of a neighnouring triangle of t)

5. If $S \neq \phi$ then obtain a new triangulation $\tilde{\tilde{\tau}}$ as follows. For all $t \in S$ (having the non-conforming point P) do. Bisect t by the longest side. Let Q be the new bisecting vertex.

   If $P \neq Q$, then join them

   $\tilde{\tau} := \tilde{\tilde{\tau}}$

   go to 4

   If $S = \phi$ then $\tau^* = \tilde{\tau}$

It is wortk pointing out here that, besides of assuring that $\tau^*$ is nested in $\tau$; the algorithm also guarantees that $\tau^*$ is conforming, non-degenerate and smooth. More details can be found in [7,8].

Examples of triangulations generated by this procedure can be seen in

Figures 2 and 3.

## 3. LOCAL SUBPROBLEMS AND SUBMSHES.

In order to discuss the adaptive multigrid algorithm, the concepts of local subproblems and submeshes need to be introduced. At this point the following assumptions are assumed to hold:

(a) A hierarchy of k conforming and nested triangulations has been generated by using iteratively an adaptive procedure and a mesh refinement procedure based on the ideas described in the preceding section.

(b) The vertices of the initial triangulation are enumerated in any order and consecutive numbers are assigned to the newly created ones.

(c) The number associated with each vertex remains fix throughout the process.

Then, the following definitions can be introduced:

Definitions. Let $N^*(\tau_k)$ be the set of the vertices of triangulation $\tau_k$ and consider $P_{k,j}$, the molecule j at level k(the polygonal region centered on node j, and whose interior only contains the vertex $j \in N^*(\tau_k)$). The vertex-set $W_{k,j}$, associated with vertex j at level k, will be the set of all the vertices of $\tau_k$ lying on the boundary of $P_{k,j}$.

If node j was created at level $m > 0$, then we shall that $W_{k,j}$ is the empty-set for $k < m$.

In addition, the triangle-set $T_{k,j}$ will be the set of all triangles of $\tau_k$ contained in $P_{k,j}$.

Figure 1 shows a typical molecule $P_{k,j}$where $W_{k,j}= \{n_1,n_2,n_3,n_4,n_5,n_6\}$ and $T_{k,j}= \{t_1,t_2,t_3,t_4,t_5,t_6\}$.

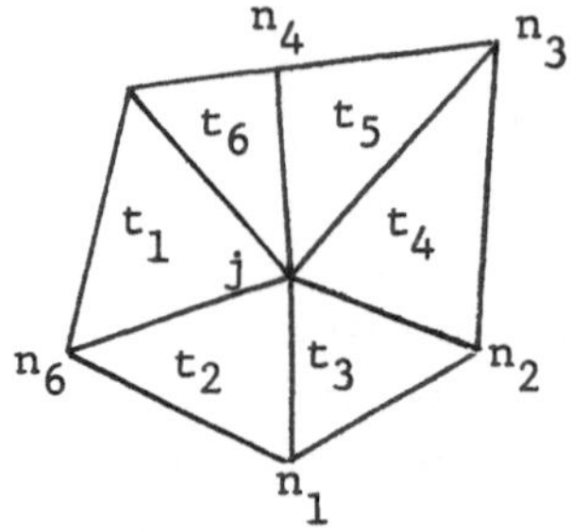


Figure 1. Molecule $P_{k,j}$, its nodes and vertices.

In order to identify the submesh associated with local refinement, the following additional definitions need to be introduced.

Definitions. Let us assume that a local refinement has been performed between levels k and k+1. Then we shall define $D_k^-$, the k submesh associated with the local refinement performed between levels k and k+1, in the following way:

$$D_k^- = \bigcup_{\substack{j \in N^*(\tau_k) \\ W_{k,j} \neq W_{k+1,j}}} T_{k,j}, \qquad D_k^- \subset \tau_k$$

Analogously, we shall define $D_{k+1}^+$, the (k+1)-submesh associated with the local refinement performed between levels k and k+1, in the following way:

$$D_{k+1}^+ = \bigcup_{\substack{j \in N^*(\tau_{k+1}) \\ W_{k+1,j} \neq W_{k,j}}} T_{k+1,j}, \qquad D_{k+1}^+ \subset \tau_{k+1}$$

In other words, the submeshes $D_k^-$ and $D_{k+1}^+$ are constructed as the union of all the triangle-sets (at their respective levels) associated with the molecules whose vertex-sets have changed between levels k and k+1. Notice that according to the definition, the vertex-sets associated with the new nodes (created between these levels) are indeed different in each level.

Fig. 2 illustrates these ideas, by taking two consecutive triangulations corresponding to levels k and k+1, where local refinement has been performed between the two levels. The local refinement at level k+1 has been drawn with dotted lines. The submeshes $D_k^-$ and $D_{k+1}^+$, respectively, are defined by the set of triangles contained in the interior of the subregion defined by the solid line. It is worth pointing out here that both submeshes cover the same subregion.

To make explicit the adaptive multigrid algorithm, we need to introduce both a parameter $\gamma$, $0 \leq \gamma < 1$, to quantify the small number of nodes created between consecutive levels, and a binary function of the level defined as follows:

$$l(k+1) = \begin{cases} 0 & \text{if } \mathrm{card}(N^*(D_{k+1}^+)) \leq \gamma \ \mathrm{card}\ (N^*(\tau_k)) \\ 1 & \text{otherwise} \end{cases}$$

where $\mathrm{card}(N^*(\tau))$ is the number of vertices of the mesh $\tau$.

In addition, it will be assumed we have at our disposal an adaptive procedure $\mathrm{adap}(\tau_k, U_k)$ and the mesh refinement procedure refine $(\tau_k, V_k)$ that together allow the construction of the hierarchy of nested triangulations.

The adaptive multigrid algorithm can be schematically described in the following way:

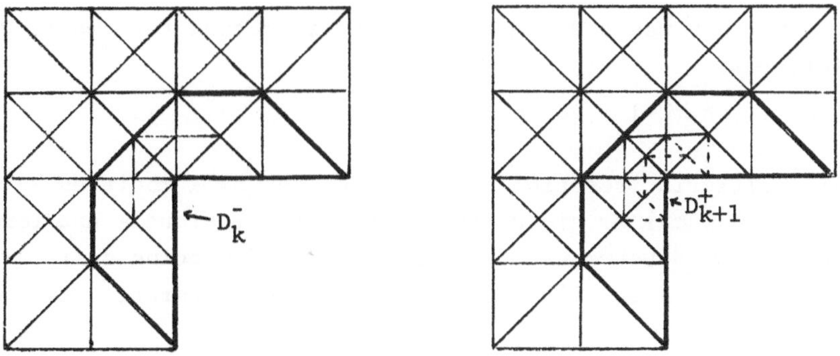


Triangulation at level k. Triangulation at level k+1

Figure 2 Representation of the submeshes $D_k^-$ and $D_{k+1}^+$ associated with local refinement performed between levels k and k+1.

Adaptive multigrid algorithm for singular solutions.

0. Input $\{\tau_o, \gamma, d\}$

1. Compute the exact finite element solution $U_o$;
   $1(0) = 1$ ; $k \leftarrow 0$;

2. Construct the submesh of refinement $V_k$
   $V_k = \text{adap}\ (\tau_k, U_k)$

3. Define the new triangulation
   $\tau_{k+1} = \text{refine}\ (\tau_k, V_k)$

4. Interpolate $U_k$ over $\tau_{k+1}$ to produce an initial estimation $\hat{U}_{k+1}$ of the finite element solution $U_{k+1}$

5. Define

$$n = \begin{cases} \text{card}\ (N^*(D^+_{k+1})) & \text{if}\ \ 1(k) = 0 \\ n + \text{card}(N^*(D^+_{k+1})) & \text{if}\ \ 1(k) = 1 \end{cases}$$

6. Is $n \leq \gamma\, \text{card}\ (N^*(\tau_k))$ ?
   a) If yes, then do d iterative sweeps over $D^+_{k+1}$ to produce an acceptable approximation $\hat{U}_{k+1}$ of $U_{k+1}$; $1(k+1) = 0$.
   b) If not, then use a dynamic multigrid iteration to compute $\hat{U}_{k+1}$; $1(k+1) = 1$.

7. $U_{k+1} \leftarrow \hat{U}_{k+1}$ ; $k \leftarrow k+1$ ; go to 2;

The dynamic multigrid iteration can be described as a modified V-cycle iteration, where local or global transfer of the residuals and iterative sweeps are performed between consecutive grids, depending on the value of the function 1. A detailed description of this iteration can be found in [10].

## 4. NUMERICAL RESULTS AND CONCLUDING REMARKS.

To illustrate the behaviour of the adaptive multigrid algorithm consider the Laplace equation with Dirichlet boundary conditions over an L-shaped region inscribed in a square of side 2 [9,10,12,13]. The solution of this problem is the function $u = r^{2/3} \sin\ (2\theta/3)$, which is singular at the origin owing to the reentrant corner. For this problem, the asymptotic rate of convergence for linear elements and quasi-uniform triangulations is 1/3.

In Table 1, the numerical results obtained with the adaptive multigrid finite element procedure has been summarized. The convergence behaviour for this problem is optimal. In effect, the adaptive procedure has achieved a numerical rate of convergence equal to 1/2. For more details see also the results published in [9,10,12,13].

Table 1

| Level | Number of unknowns | Number of triangles | Percentual energy norm error |
|---|---|---|---|
| 1 | 12 | 38 | 17.8 |
| 3 | 16 | 46 | 14.2 |
| 5 | 26 | 68 | 11.9 |
| 8 | 40 | 98 | 9.5 |
| 11 | 57 | 134 | 7.8 |
| 12 | 76 | 176 | 7.1 |
| 14 | 104 | 240 | 6.1 |
| 17 | 169 | 380 | 4.8 |
| 20 | 263 | 574 | 3.9 |
| 23 | 412 | 888 | 3.1 |
| 24 | 465 | 1000 | 2.9 |
| 26 | 657 | 1396 | 2.5 |

Table 1 only includes the results corresponding to the levels with $\ell(k) = 1$, namely those levels where a dynamic V-cycle iteration was performed to obtain the finite element solution. In Figure 3 we have included the triangulations adaptively constructed corresponding to levels 12,14,17 and 20.

Finally, to conclude it should be pointed out that new versions of the mesh refinement algorithms of Rivara [14] have been recently developed in such a way that each one of the triangles of the refinement submesh V is refined in 4 parts. By replacing the algorithms described in the second section by these new algorithms, we have succeeded to reduce the number of triangulations generated, and consequentely the computational work involved.

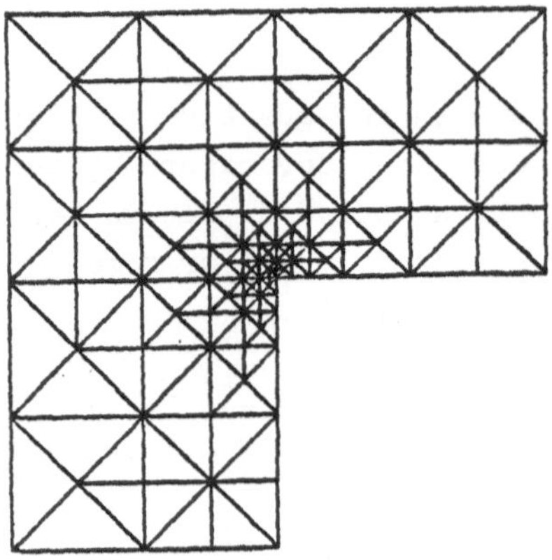

level 12

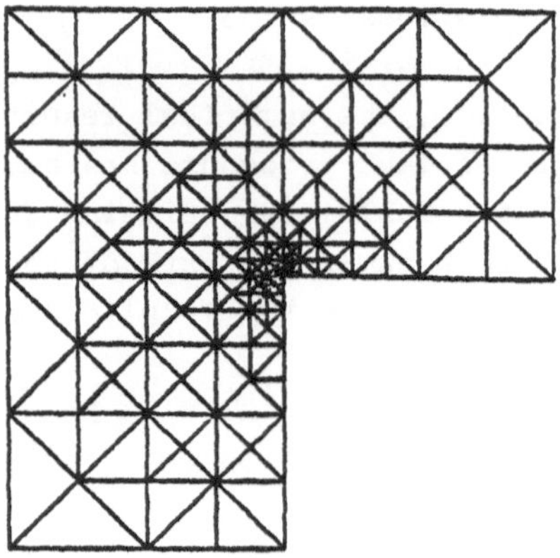

level 14

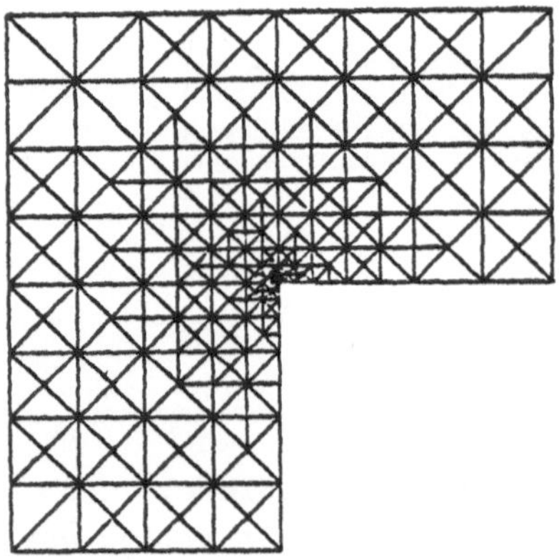

level 17

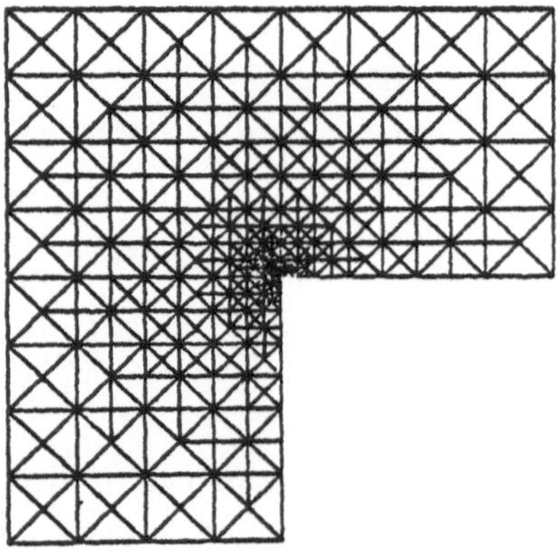

level 20

Figure 3

Examples of triangulations adaptively constructed for the L-shaped problem.

ACKNOWLEDGEMENTS.

This work has been supported in part by Proyecto 1122/84 Fondo Nacional de Desarrollo Científico y Tecnológico de Chile.

REFERENCES.

[1] Babuska, I. The selfadaptive approach in the finite element method.,pp. 125-142 of J.R. Whiteman (Ed.),The Mathematics of Finite Elements and Applications. Academic Press, London (1976).

[2] Babuska, I. and Rheinboldt,W.C., Error estimates for adaptive finite element computations, SIAM J. Numer. Anal. 15, 736-754 (1978).

[3] Babuska, I. and Rheinboldt,W.C., Reliable error estimation and mesh adaptation for the finite element method,pp.67-108 of J.T. Oden (ed.), Computational Methods in Nonlinear Mechanics, North Holland (1980).

[4] Babuska,I. and Rheinboldt,W.C., A survey of a-posteriori error estimators and adaptive approaches in the finite element method. Institute for Physical Sciences and Technology, University of Maryland. Technical Note BN 981, 1982.

[5] Hackbusch,W. and U.Trottenberg (eds.) Multigrid Methods,Lecture Notes in Math. 960, Springer Verlag, 1982.

[6] Bank,R.E. and Sherman A.H., The use of adaptive grid refinement for badly behaved elliptic partial differential equations,pp. 18-24 of Computers in Simulation XXII, North Holland (1980).

[7] Rivara,M.C., Algorithms for refining triangular grids suitable for adaptive and multigrid techniques, Int.J.Numer. Meth. Engrg. 20,pp.745-756 (1984).

[8] Rivara,M.C., Mesh refinement processes based on the generalized bisection of simplices, SIAM J.on Numer.Anal., 21,604-613 (1984).

[9] Rivara,M.C., Design and data structure of a fully adaptive multigrid finite element software, ACM Trans. on Math. Software 10,242-264 (1984).

[10] Rivara, M.C., Adaptive multigrid software for the finite element method, Doctoral dissertation, K.U.Leuven, Belgium (1984).

[11] Rivara, M.C., EXPDES User's Manual, Dept. Computer Science, K.U.Leuven (1984).

[12] Rivara, M.C., Dynamic implementation of the h-version of the finite element method, in The Mathematics of Finite Elements and Applications, Brunel University, (1984).

[13] Rivara,M.C., Adaptive finite element refinement and fully irregular and conforming triangulations. To appear in Accuracy Estimates and Adaptive Refinements in Finite Element Computations, Babuska, Oliveira and Zienkiewicz (eds.) John Wiley and Sons Pub.

[14] Rivara,M.C., A grid generator based on 4-triangles mesh-refinement algorithms, Dept. of Mathematics and Computer Sciences University of Chile, 1985.

# Lecture Notes in Control and Information Sciences

Edited by M. Thoma

Vol. 43: Stochastic Differential Systems
Proceedings of the 2nd Bad Honnef Conference of the SFB 72 of the DFG at the University of Bonn
June 28 – July 2, 1982
Edited by M. Kohlmann and N. Christopeit
XII, 377 pages. 1982.

Vol. 44: Analysis and Optimization of Systems
Proceedings of the Fifth International Conference on Analysis and Optimization of Systems
Versailles. December 14-17, 1982
Edited by A. Bensoussan and J. L. Lions
XV, 987 pages, 1982

Vol. 45: M. Arató
Linear Stochastic Systems with Constant Coefficients
A Statistical Approach
IX, 309 pages. 1982

Vol. 46: Time-Scale Modeling of Dynamic Networks with Applications to Power Systems
Edited by J. H. Chow
X, 218 pages. 1982

Vol. 47: P. A. Ioannou, P. V. Kokotovic
Adaptive Systems with Reduced Models
V, 162 pages. 1983

Vol. 48: Yaakov Yavin
Feedback Strategies for Partially Observable Stochastic Systems
VI, 233 pages, 1983

Vol. 49: Theory and Application of Random Fields
Proceedings of the IFIP-WG 7/1 Working Conference held under the joint auspices of the Indian Statistical Institute
Bangalore, India, January 1982
Edited by G. Kallianpur
VI. 290 pages. 1983

Vol. 50: M. Papageorgiou
Applications of Automatic Control Concepts to Traffic Flow Modeling and Control
IX, 186 pages. 1983

Vol. 51: Z. Nahorski, H.F. Ravn, R.V.V. Vidal
Optimization of Discrete Time Systems
The Upper Boundary Approach
V, 137 pages 1983

Vol. 52: A. L. Dontchev
Perturbations, Approximations and Sensitivity Analysis of Optimal Control Systems
IV, 158 pages. 1983

Vol. 53: Liu-Chen Hui
General Decoupling Theory of Multivariable Process Control Systems
XI, 474 pages. 1983

Vol. 54: Control Theory for Distributed Parameter Systems and Applications
Edited by F. Kappel, K. Kunisch, W. Schappacher
VII, 245 pages. 1983.

Vol. 55: Ganti Prasada Rao
Piecewise Constant Orthogonal Functions and Their Application to Systems and Control
VII, 254 pages. 1983.

Vol. 56: Dines Chandra Saha, Ganti Prasada Rao
Identification of Continuous Dynamical Systems
The Poisson Moment Functional (PMF) Approach
IX, 158 pages. 1983.

Vol. 57: T. Söderström, P. G. Stoica
Instrumental Variable Methods for System Identification
VII, 243 pages. 1983.

Vol. 58: Mathematical Theory of Networks and Systems
Proceedings of the MTNS-83 International Symposium
Beer Sheva, Israel, June 20–24, 1983
Edited by P. A. Fuhrmann
X, 906 pages. 1984

Vol. 59: System Modelling and Optimization
Proceedings of the 11th IFIP Conference
Copenhagen, Denmark, July 25-29, 1983
Edited by P. Thoft-Christensen
IX, 892 pages. 1984

Vol. 60: Modelling and Performance Evaluation Methodology
Proceedings of the International Seminar
Paris, France, January 24–26, 1983
Edited by F. Bacelli and G. Fayolle
VII, 655 pages. 1984

Vol. 61: Filtering and Control of Random Processes
Proceedings of the E.N.S.T.-C.N.E.T. Colloquium
Paris, France, February 23–24, 1983
Edited by H. Korezlioglu, G. Mazziotto, and J. Szpirglas
V, 325 pages. 1984

# Lecture Notes in Control and Information Sciences

Edited by M. Thoma and A. Wyner

Vol. 62: Analysis and Optimization of Systems
Proceedings of the Sixth International Conference on Analysis and Optimization of Systems
Nice, June 19–22, 1984
Edited by A. Bensoussan, J. L. Lions
XIX, 591 pages. 1984.

Vol. 63: Analysis and Optimization of Systems
Proceedings of the Sixth International Conference on Analysis and Optimization of Systems
Nice, June 19–22, 1984
Edited by A. Bensoussan, J. L. Lions
XIX, 700 pages. 1984.

Vol. 64: Arunabha Bagchi
Stackelberg Differential Games in Economic Models
VIII, 203 pages, 1984

Vol. 65: Yaakov Yavin
Numerical Studies in Nonlinear Filtering
VIII, 273 pages, 1985.

Vol. 66: Systems and Optimization
Proceedings of the Twente Workshop
Enschede, The Netherlands, April 16–18, 1984
Edited by A. Bagchi, H. Th. Jongen
X, 206 pages, 1985.

Vol. 67: Real Time Control of Large Scale Systems
Proceedings of the First European Workshop
University of Patras, Greece, Juli 9–12, 1984
Edited by G. Schmidt, M. Singh, A. Titli, S. Tzafestas
XI, 650 pages, 1985.

Vol. 68: T. Kaczorek
Two-Dimensional Linear Systems
IX, 397 pages, 1985.

Vol. 69: Stochastic Differential Systems – Filtering and Control
Proceedings of the IFIP-WG 7/1 Working Conference
Marseille-Luminy, France, March 12-17, 1984
Edited by M. Metivier, E. Pardoux
X, 310 pages, 1985.

Vol. 70: Uncertainty and Control
Proceedings of a DFVLR International Colloquium
Bonn, Germany, March, 1985
Edited by J. Ackermann
IV, 236 pages, 1985.

Vol. 71: N. Baba
New Topics in Learning Automata Theory and Applications
VII, 231 pages, 1985.

Vol. 72: A. Isidori
Nonlinear Control Systems: An Introduction
VI, 297 pages, 1985.

Vol. 73: J. Zarzycki
Nonlinear Prediction Ladder-Filters for Higher-Order Stochastic Sequences
V, 132 pages, 1985.

Vol. 74: K. Ichikawa
Control System Design based on Exact Model Matching Techniques
VII, 129 pages, 1985.

Vol. 75: Distributed Parameter Systems
Proceedings of the 2nd International Conference, Vorau, Austria 1984
Edited by F. Kappel, K. Kunisch, W. Schappacher
VIII, 460 pages, 1985.

Vol. 76: Stochastic Programming
Edited by F. Archetti, G. Di Pillo, M. Lucertini
V, 285 pages, 1986.

Vol. 77: Detection of Abrupt Changes in Signals and Dynamical Systems
Edited by M. Basseville, A. Benveniste
X, 373 pages, 1986.

Vol. 78: Stochastic Differential Systems
Proceedings of the 3rd Bad Honnef Conference, June 3–7, 1985
Edited by N. Christopeit, K. Helmes, M. Kohlmann
V, 372 pages, 1986.

Vol. 79: Signal Processing for Control
Edited by K. Godfrey, P. Jones
XVIII, 413 pages, 1986.

Vol. 80: Artificial Intelligence and Man-Machine Systems
Edited by H. Winter
IV, 211 pages, 1986.

# Lecture Notes in Control and Information Sciences

Edited by M. Thoma and A. Wyner

Vol. 81: Stochastic Optimization
Proceedings of the International Conference,
Kiew, 1984
Edited by I. Arkin, A. Shiraev, R. Wets
X, 754 pages, 1986.

Vol. 82: Analysis and Algorithms
of Optimization Problems
Edited by K. Malanowski, K. Mizukami
VIII, 240 pages, 1986.

Vol. 83: Analysis and Optimization
of Systems
Proceedings of the Seventh International
Conference of Analysis and Optimization
of Systems
Antibes, June 26-27, 1986
Edited by A. Benoussan, J. L. Lions
XVI, 901 pages, 1986.

Vol. 84: System Modelling
and Optimization
Proceedings of the 12th IFIP Conference
Budapest, Hungary, September 2–6, 1985
Edited by A. Prékopa, J. Szelezsán, B. Strazicky
XII, 1046 pages, 1986.

Vol. 85: Stochastic Processes
in Underwater Acoustics
Edited by Charles R. Baker
V, 205 pages, 1986.

Vol. 86: Time Series and
Linear Systems
Edited by Sergio Bittanti
XVII, 243 pages, 1986.

Vol. 87: Recent Advances in
System Modelling and Optimization
Proceedings of the IFIP-WG 7/1
Working Conference
Santiago, Chile, August 27-31, 1984
Edited by L. Contesse, R. Correa, A. Weintraub
IV, 199 pages, 1987.